2025 International Conference on Advanced Sensing Technology and Aerospace (ASTA 2025)

Journal of Physics: Conference Series Volume 3056

Changsha, China
25-28 April 2025

ISBN: 979-8-3313-2481-0
ISSN: 1742-6588

Printed from e-media with permission by:

Curran Associates, Inc.
57 Morehouse Lane
Red Hook, NY 12571

Some format issues inherent in the e-media version may also appear in this print version.

Institute of Physics
Dirac House, Temple Back
Bristol BS1 6BE UK

Phone: 44 1 17 929 7481
Fax: 44 1 17 920 0979

techtracking@iop.org

Additional copies of this publication are available from:

Curran Associates, Inc.
57 Morehouse Lane
Red Hook, NY 12571 USA
Phone: 845-758-0400
Fax: 845-758-2633
Email: curran@proceedings.com
Web: www.proceedings.com

TABLE OF CONTENTS

Preface

It is our great pleasure to present the proceedings of the 2025 International Conference on Advanced Sensing Technology and Aerospace (ASTA2025). This conference serves as a premier platform for researchers, engineers, and practitioners from around the world to share their latest findings, innovations, and insights in the fields of advanced sensing technologies and aerospace engineering.

We are deeply grateful to our distinguished keynote speakers, Prof. Dongzhi Zhang, Prof. Pavel Loskot and invited speaker Prof. Xiang Zhang, whose inspiring presentations have set the tone for this year’ s conference and provided invaluable perspectives that have enriched the discussions throughout the event. Their expertise and visionary outlook continue to drive progress and innovation in our community.

We also extend our sincere thanks to the guest editors Prof. Jin Zhu and Zhenpeng He, for their dedicated efforts in reviewing and selecting high-quality papers. Their rigorous evaluation and constructive feedback have ensured the excellence and relevance of the contributions included in these proceedings.

Special appreciation goes to the organizing committee and all supporting institutions(Hunan University of Technology, Central South University) whose hard work and meticulous planning have made ASTA2025 a successful and memorable event. Their commitment to fostering collaboration and knowledge exchange is the cornerstone of this conference’s ongoing success.

We sincerely thanks every participant who has contributed to making ASTA2025 a remarkable gathering of minds and hope that the research presented in these proceedings will inspire further advancements and collaborations in the exciting fields of sensing technology and aerospace, and contribute to the development of innovative solutions for global challenges.

Journal of Physics: Conference Series **3056** (2025) 011001 doi:10.1088/1742-6596/3056/1/011001

COMMITTEES

General Chairs
Dongzhi Zhang, China University of Petroleum (East China), China
Shibin Luo,Central South University, China

Program Chairs
Shouwen Liu, Beijing Satellite Environmental Engineering Institute, China
Daren Zheng, Nanjing University of Aeronautics and Astronautics, China
Xing Sun, Northwestern Polytechnical University, China

Technical Chairs
Yuchen Jiang, Harbin Institute of Technology, China
Zhenpeng He, Civil Aviation University of China, China
Jin Zhu, Jiangsu University of Science and Technology, China

Publicity Chairs
Yuqiang Yang, Guangdong Ocean University, China
Ke Li, Beihang University, China
Xiang Zhang, Harbin Institute of Technology, China

Publication Chair
Ji Wang, Guangdong Ocean University, China
Yuanfa Ji, Guilin University Of Electronic Technology, China
Xianming Huang, Hunan University of Technology, China

Technical Program Committees
Wen Cai, Northwestern Polytechnical University, China
Haichao An, Beijing Institute of Technology, China
Liang Zhang, Beihang University, China
Chen Yang, Beihang University, China
Dongsoo Har, Korea Advanced Institute of Science and Technology, South Korea
Mumtaz Karatas, Wright State University, USA
Chao An, Beihang University, China
Poonam Rani, NSUT Main Campus, India
Jianxun Zhang, Xi 'an Jiaotong University, China
Xi Xu, Northwestern Polytechnical University, China
Yimeng Ge, Xi’ an University of Posts & Telecommunications, China
Chun Liu, Shanghai University, China
Feng Zhang, Northwestern Polytechnical University, China
Xueguang Meng, Xi'an Jiaotong University, China
Ayesha Abbas, NED university of Engineering and Technology, Pakistan
Jihe Wang, Northwestern Polytechnical University, China
Jiangtao Wang, Beijing Institute of Technology, China

Journal of Physics: Conference Series **3056** (2025) 011001 doi:10.1088/1742-6596/3056/1/011001

Zhengze Zhang, Northwestern Polytechnical University, China
Jun Dong, Xi 'an Jiaotong University, China
Samet Diri, Kocaeli University, Turkey

Peer Review Statement

All papers published in this volume have been reviewed through processes administered by the Editors. Reviews were conducted by expert referees to the professional and scientific standards expected of a proceedings journal published by IOP Publishing.

- **Type of peer review:** Single Anonymous
- **Conference submission management system:** Morressier
- **Number of submissions received:** 48
- **Number of submissions sent for review:** 48
- **Number of submissions accepted:** 35
- **Acceptance Rate (Submissions Accepted / Submissions Received × 100):** 72.9
- **Average number of reviews per paper:** 1
- **Total number of reviewers involved:** 48
- **Contact person for queries:**
  Name: Zhenpeng He
  Email: asta2025@126.com
  Affiliation: Civil Aviation University of China

Interval Type-2 Fuzzy PID Control for Space Manipulator with Flexible Links and Joints

Mingyu Li[1],Menglu Zhu[1],Chenhao Liu[1],Hongzhen Li[1],Yi Zeng*[1]
1. School of Astronautics, Harbin Institute of Technology, Harbin, China
* E-mail: yi-zeng1991@outlook.com

Abstract. The space industry has made remarkable progress in recent years. Space exploration is deepening and space tasks are becoming increasingly complex. In order to meet the higher requirements for space exploration equipment, this paper designs a fuzzy PID space manipulator tracking control method. Aiming at the planar two flexible-link and flexible-joint space manipulator tracking control task, the non-singleton interval type-2 fuzzy PID (N-IT2FPID) controller is designed. A refined version of the particle swarm optimization (PSO) technique is introduced to enhance fuzzy control system. The simulation outcomes indicate that the PSO-optimized fuzzy PID controller demonstrates superior performance by comparison. Numberical experimental validation confirms that the introduced controller is capable of efficiently managing the flexible space manipulator.
Keywords: space manipulator, interval type-2 fuzzy control, fuzzy PID, PSO

1. Introduction

In the process of space exploration, astronauts need to overcome the difficulties of space radiation, microgravity environment, extreme temperature and so on in order to carry out space missions, which not only reduces the efficiency of the mission, but also seriously threatens the safety of astronauts. In recent years, space robots are more and more widely used. Space robot is a special robot used to assist or replace human beings in the space environment for extravehicular operations, space exploration and other activities. The space manipulator is a special kind of space robot, which is similar to the human arm, and usually consists of a base and several rods. The space manipulator has strong capture, rendezvous and docking capabilities as well as loading capacity, and has become essential equipment for construction, upkeep, and operation of space stations [1].

Space manipulator controller is very important for space manipulator. It is directly related to the performance, accuracy and success rate of space manipulator missions. The efficient control algorithm can guarantee precise and stable manipulation control, optimizing both the trajectory and force distribution of the space manipulator, reduce energy consumption, reduce task cost. Since IT2F logic was introduced into PID control, IT2FPID controller has been widely used for manipulators [2]. The controller combines flexibility of former and accuracy latter, and can adjust adaptively according to the real-time environmental information, so as to improve the control effect. For IT2FPID controller, non-singleton fuzzification [3] can better deal with uncertainty than traditional singleton fuzzification. An pre compensated IT2FPID controller is proposed in [4] for tracing a variable load two degree of freedom manipulator. In [5], an fractional-order FPID controller is developed for manipulator. Real-time simulation results have shown that the introduced controller exhibits greater robustness against external disturbances and more effectively minimizes the impact of uncertainties compared to PID and traditional FPID controllers. A metaheuristic most valuable player algorithm (MVPA) is used to tune FPID controllers' parameters in [6]. The robustness of these controllers has been explored for initial conditions, disturbance rejection, and model uncertainty.

Fuzzy control algorithm can also be combined with other intelligent algorithms to further optimize the control performance. PSO is a swarm intelligence optimization algorithm, which has been widely used. In [7], a FPID controller, which utilizes self-tuning capabilities through PSO, is crafted on the foundation of T1FPID control. The PSO algorithm optimizes the quantization and scaling factors within the fuzzy logic, aiming to attain both high robustness and precision for rigid manipulator. In [8], an advanced method for tracking electric vehicle driven by brushless DC motor using PSO is presented. This approach aims to fine-tune PID gains in coordination with a fuzzy controller.

Contributions of this article are as follows:

- It is difficult to design controllers because of highly nonlinearity and complexity for space manipulators, so there is little research on fuzzy PID control of this kind of such flexible space manipulator. In this paper, a N-IT2FPID controller is proposed for a given flexible-link and flexible-joint space manipulator model. The proposed controller combines the advantages of IT2F logic and traditional PID controller. It not only possesses the capability to address the uncertainty inherent in complex systems, but also has strong robustness and flexibility. The numerical experiment demonstrates that proposed controller outperforms both traditional PID controller and S-IT2FPID controller.
- For control system, traditional PSO optimization algorithm has some problems, such as the limitation of lacking global search capability in the initial stage, and the fitness is difficult to reflect the system control performance. In order to solve these problems, PSO algorithm is improved in this paper. According to the characteristics of the plant with multiple inputs and multiple outputs (MIMO), a new PSO fitness calculation method is proposed to take into account the accuracy and rapidity of the system.

First, a space manipulator dynamic model with flexible links and joints is given in section 2. In section 3, N-IT2FPID controller is designed for the manipulator tracking control task. In section 4, improved PSO method is used to optimize the fuzzy system. The numerical experiment in section 5 show that the PSO optimized FPID controller can meet the basic requirements of flexible space manipulator control, and PSO algorithm can optimize the FPID controller and improve the control effect.

2. Flexible-link and flexible-joint space manipulator model

It is assumed that the space manipulator is not affected by external forces such as gravity, can be modeled in the inertial coordinate system, and all joints are flexible and rotatable. All links are flexible. In this paper as shown in figure 1, a planar two-joint space manipulator with flexible links and joints [9] is modeled as the plant. The Lagrange equations can be utilized to deduce dynamic model of the space manipulator. More details are shown in [9].

In figure 1,$\{\mathrm{I}\}$ is inertial coordinate system, $\mathrm{C}_1, \mathrm{C}_2$ are centroids of links, C_b is centroid of base, $\mathrm{J}_1, \mathrm{J}_2$ are two joints, $\boldsymbol{r}_b$ is base's position vector, θ_0 is posture of base, θ_1, θ_2 are joint angles.

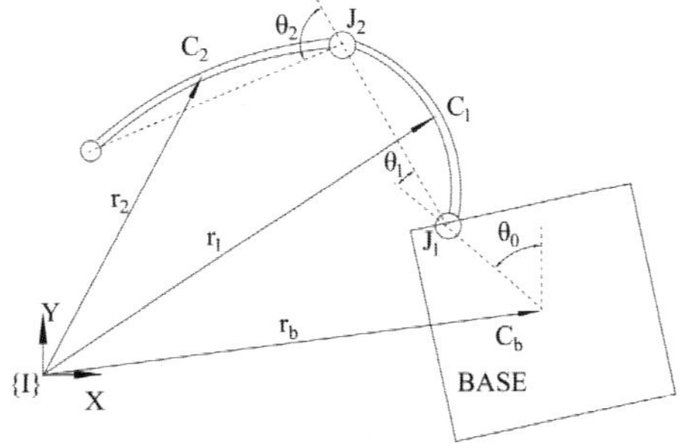

Figure 1: planar two link flexible space manipulator

The model is presented in [9]

$$\begin{aligned} \mathrm{D}(\boldsymbol{q})\ddot{\boldsymbol{q}} + \mathrm{C}(\boldsymbol{q},\dot{\boldsymbol{q}})\dot{\boldsymbol{q}} + \begin{bmatrix} \mathbf{0}_{5\times1} \\ \mathrm{K}_\delta \boldsymbol{\delta} \end{bmatrix} &= \begin{bmatrix} \boldsymbol{F} \\ \tau_0 \\ \boldsymbol{\tau} \\ \mathbf{0}_{4\times1} \end{bmatrix}, \\ J_m \ddot{\boldsymbol{q}}_m + \boldsymbol{\tau} &= \boldsymbol{\tau}_m, \\ \boldsymbol{\tau} &= \mathrm{K}_m \boldsymbol{\sigma}, \end{aligned} \tag{1}$$

where $\boldsymbol{q} = [\boldsymbol{r}_b^{\mathrm{T}}, \theta_0, \theta_1, \theta_2, \boldsymbol{\delta}^{\mathrm{T}}]^{\mathrm{T}}$ is generalized coordinate, $\boldsymbol{\delta}$ is modal coordinate vector of flexible rod, $\mathrm{D}(\boldsymbol{q})$ is positive definite inertia matrix, $\mathrm{C}(\boldsymbol{q},\dot{\boldsymbol{q}})\dot{\boldsymbol{q}}$ is centrifugal force and Coriolis force vector, $\mathrm{K}_\delta\boldsymbol{\delta}$ is equivalent stiffness matrix of rod, $\boldsymbol{F}$ is position control force of the base, τ_0 is the torque on base, $\boldsymbol{\tau}_m$ is the rotor control torque of motor, J_m is rotor moment of inertia matrix of motor, $\boldsymbol{\sigma} = \boldsymbol{q}_m - \boldsymbol{q}_j$, $\boldsymbol{q}_m = [\theta_{m1}, \theta_{m2}]$ is the rotor angle of the motor, $\boldsymbol{q}_j = [\theta_1, \theta_2]$, and K_m is the equivalent stiffness matrix of the flexible joint.

3. N-IT2FPID controller design

Different from S-IT2FPID controller, this paper adopts the N-IT2FPID controller which is more suitable for dealing with uncertainty. The input signal of IT2F logic system is error signal e_s and its differential signal e_{ds}. The paper selects five fuzzy subsets and adopts the membership function of the bottom width triangle, as follows [3]:

$$\underline{\mu}_{Ai}(x) = \begin{cases} -\frac{x-x_i}{a}+1 & \text{, if } x_i \le x \le x_i + a, \\ \frac{x-x_i}{a}+1 & \text{, if } x_i - a \le x \le x_i, \\ 0 & \text{, else,} \end{cases} \tag{2a}$$

$$\bar{\mu}_{Ai}(x) = \begin{cases} -\frac{x-x_i}{b}+1 & \text{, if } x_i \le x \le x_i + b, \\ \frac{x-x_i}{b}+1 & \text{, if } x_i - b \le x \le x_i, \\ 0 & \text{, else,} \end{cases} \tag{2b}$$

where $b > a > 0, i = 1, 2, 3, 4, 5$.

According to the fuzzy rules, fuzzy inference is carried out on the fuzzy value. Finally, the fuzzy inference result is made clear by using the corresponding algorithm for $\Delta K_{P2i}, \Delta K_{I2i}, \Delta K_{D2i}$ .Where the controller expression is given in (3) [10]:

$$u_i = (K_{Pi} + \Delta K_{P2i})e_i + (K_{Ii} + \Delta K_{I2i}) \int_0^{\mathrm{T}} e_i \mathrm{d}t + (K_{Di} + \Delta K_{D2i})\dot{e}_i, \tag{3}$$

where, $\Delta K_{P2i}, \Delta K_{I2i}, \Delta K_{D2i}$ are output by IT2F logic system, T is upper bound of integral. See table 1 for basic parameters of PID controllers, 5 IT2F subsets output ranges in table 2, fuzzy rules in table 3,4,5.

Table 1: Basic Parameters Of PID Controllers

controller i	K_{Pi}	K_{Ii}	K_{Di}	controller output
1	10	0.5	1.2	τ_{m1}
2	7	0.5	1	τ_{m2}

Table 2: IT2F Subsets Output Ranges

	NB	NS	Z	PS	PB
$\bar{y}^j$	-0.9	-0.4	0.1	0.6	1
$\underline{y}^j$	-1	-0.6	-0.1	0.4	0.9

Table 3: KP fuzzy rules

$e_s \backslash e_{ds}$	NB	NS	Z	PS	PB
NB	PS	PS	NS	PS	PB
NS	PS	Z	Z	Z	Z
Z	NS	Z	Z	Z	NS
PS	Z	Z	Z	Z	PS
PB	PB	PS	NS	PS	PS

Table 4: KI fuzzy rules

$e_s \backslash e_{ds}$	NB	NS	Z	PS	PB
NB	NB	NS	NS	NS	Z
NS	NB	NS	NS	Z	PS
Z	NS	NS	Z	PS	PS
PS	NS	Z	PS	PS	PB
PB	Z	PS	PS	PS	PB

Table 5: KD fuzzy rules

$e_s \backslash e_{ds}$	NB	NS	Z	PS	PB
NB	PB	NS	NS	NS	Z
NS	PS	Z	Z	Z	Z
Z	PS	PS	Z	PS	PS
PS	Z	Z	Z	Z	PS
PB	Z	NS	NS	NS	PB

reduction and defuzzification algorithms are as below, Let $\boldsymbol{x} = (x_1, x_2, \ldots, x_p)$ be the clear value input of fuzzy system. A total of N fuzzy rules are set. After fuzzy reasoning, the fire interval of the j th fuzzy rule is obtained as [11]

$$\mathrm{F}^j(x) = [\underline{f}^j, \bar{f}^j]. \tag{4}$$

The output value of the j th fuzzy rule y_j is

$$y_j = [\underline{y}^j, \bar{y}^j], \tag{5}$$

where

$$\underline{f}^j = [\underline{\mu}_{\widetilde{A}_1^j}(x_1) \times \cdots \times \underline{\mu}_{\widetilde{A}_p^j}(x_p)], \bar{f}^j = [\bar{\mu}_{\widetilde{A}_1^j}(x_1) \times \cdots \times \bar{\mu}_{\widetilde{A}_p^j}(x_p)]. \tag{6}$$

Next, use the center algorithm [10] to reduce the type as shown in (7a) and (7b):

$$y_l = \frac{\sum_{j=1}^{L} \underline{y}^j \bar{f}^j + \sum_{j=L+1}^{N} \underline{y}^j \underline{f}^j}{\sum_{j=1}^{L} \bar{f}^j + \sum_{j=L+1}^{N} \underline{f}^j}, y_u = \frac{\sum_{j=1}^{U} \bar{y}^j \underline{f}^j + \sum_{j=U+1}^{N} \bar{y}^j \bar{f}^j}{\sum_{j=1}^{U} \underline{f}^j + \sum_{j=U+1}^{N} \bar{f}^j}. \tag{7}$$

In (7), the switch point L can be obtained by enhanced iterative algorithm with stop condition (EIASC) [11] :

Step 1 Sort $\underline{y}^j$ in ascending order $\underline{y}^j, j = 1, \ldots, n$.

Step 2 Initialize a, b, L

$$a = \sum_{n=1}^{N} \underline{y}^n \underline{f}^n, b = \sum_{n=1}^{N} \underline{f}^n, L = 0. \tag{8}$$

Step 3 Compute a, b, L, y_l

$$L = L + 1, a = a + \underline{y}^L \left(\bar{f}^L - \underline{f}^L\right), b = b + \bar{f}^L - \underline{f}^L, y_l = a/b. \tag{9}$$

Step 4 If $y_l \leq \underline{y}^{L+1}$, output L; else, go back to Step 3, compute new a, b, L, y_l.

The switch point U can be calculated in a similar way [11].

The average value of the two in (7) y is the clear output.

$$y = \frac{y_l + y_u}{2}. \tag{10}$$

4. PSO for N-IT2FPID controller

In this paper, PSO is employed to fine-tune the parameters of IT2FPID controller. The specific optimized parameters are fuzzy scale factors and quantization factors in fuzzy PID controllers.

Firstly, multiple particles are generated within a certain range. Each particle has two basic attributes: position and speed. Its initial position and speed are randomly generated, and then the iterative search for the optimal solution is started until the search target is reached [12]:

$$\boldsymbol{v}_i^{k+1} = \omega \cdot \boldsymbol{v}_i^k + c_1 \cdot \mathrm{R}_1 \cdot \left(\boldsymbol{x}_{lbi}^k - \boldsymbol{x}_i^k\right) + c_2 \cdot \mathrm{R}_2 \cdot \left(\boldsymbol{x}_{gbi}^k - \boldsymbol{x}_i^k\right), \tag{11a}$$

$$\boldsymbol{x}_i^{k+1} = \boldsymbol{x}_i^k + \boldsymbol{v}_i^{k+1}, \tag{11b}$$

where $\boldsymbol{x}_i^k = [\boldsymbol{g}_{oi}^{k\mathrm{T}}, \boldsymbol{g}_{si}^{k\mathrm{T}}]^\mathrm{T}$ is particle position i in iter k while $\boldsymbol{x}_i^{k+1}$ in iter $k+1$, $\boldsymbol{g}_{oi6\times1}^k$ is optimized parameter vector, $\boldsymbol{g}_{si4\times1}^k$ is scale parameter vector. ω is the inertia factor. $\boldsymbol{x}_{lbi}^k$ is the optimal position of the particle itself, and $\boldsymbol{x}_{gbi}^k$ is the optimal position of all particles obtained in the iteration process. $\mathrm{R}_1, \mathrm{R}_2$ are random number matrices. The matrix elements are in the (0,1) interval. c_1, c_2 are learning factors, representing influence of particle's own experience and global experience on velocity.

The inertia coefficient of the traditional PSO algorithm is a fixed value, which is improved by this paper to decay with the number of iterations, in order to enhance the algorithm's global search capability during initial stage and its local search ability in later stage, as shown in following formula:

$$\omega = \omega_{\min} + (\omega_{\max} - \omega_{\min}) \cdot \left[1 - \left(\frac{i}{i_{\max}}\right)^3\right], \tag{12}$$

where, i is number of iterations, $\omega_{\min}$ is minimum inertia coefficient, $\omega_{\max}$ is maximum inertia coefficient.

Particle fitness attribute is calculated from the particle position. If for iteration k, the fitness function of a particle is the smallest of all particles, then the particle position is called the global optimal position of this iteration, that is, $\boldsymbol{x}_{gbi}^{k}$ in (11a). The goal of PSO is to reduce the fitness of the global optimal location [12]. For traditional particle swarm optimization, the performance evaluation index ITAE [13] of the control system is generally equal to the particle fitness J. Small ITAE usually shows good performance of the control system. In (13), $|e(t)|$ is the absolute value of error.

$$\mathrm{ITAE} = \int_0^{\mathrm{T}} t|e(t)|\mathrm{d}t. \tag{13}$$

This fitness calculation method helps to reduce the system adjustment time and overshoot, but when t is small, the error has little impact on ITAE, resulting in the bad rapidity of the system. Therefore, according to the MIMO characteristics of manipulator, taking into account the accuracy and rapidity of the system, a new fitness J calculation method is proposed in this paper

$$J = \int_0^{T_1} t\left(\alpha_1 \cdot |e_1(t)| + \alpha_2 \cdot |e_2(t)|\right)\mathrm{d}t + \int_0^{T_2} \left(\beta_1 \cdot |e_1(t)| + \beta_2 \cdot |e_2(t)|\right)\mathrm{d}t, \tag{14}$$

where, $\alpha_1, \alpha_2, \beta_1, \beta_2$ are the weighting factors, $T_2 \leq T_1$, and $|e_1(t)|, |e_2(t)|$ are the absolute values of the angle errors of joint 1,2.

5. Numerical experiment

In order to verify and demonstrate the effectiveness of the improved PSO-optimized N-IT2FPID controller in the paper, the following numerical experiment are carried out. The parameters of IT2F membership function a=0.25, b=0.5, x_i in (2a),(2b) are shown in the table 6.

Table 6: Membership function parameter x_i

	NB	NS	Z	PS	PB
x_i	-1	-0.5	0	0.5	1

Using the designed controller, give the input signal $\boldsymbol{q}_0 = [\boldsymbol{r}_b^{\mathrm{T}}, \theta_0, \theta_1, \theta_2]^{\mathrm{T}} = [\mathbf{0}, -1, 1, 1]^{\mathrm{T}}$, and compare the output signal, as shown in figure 2,3. The errors are shown in figure 4,5. The control performance indicators are shown in the table 7. "S" and "N" represent traditional singleton fuzzification and non-singleton fuzzification, respectively. Small IAE [13] in (15) also shows good system performance.

$$\mathrm{IAE} = \int_0^{\mathrm{T}} |e(t)|\mathrm{d}t. \tag{15}$$

From the output signal, performance of IT2FPID controller without PSO is relatively poor, and the overshoot is large, which is difficult to meet the stability requirements. However, the improved PSO algorithm for IT2FPID controller improves this shortcoming, making the system meet the task requirements. In table 7, the performance index of non-singleton fuzzification is better than singleton fuzzification.

Then give the input signal $\boldsymbol{q}_0 = [\boldsymbol{r}_b^{\mathrm{T}}, \theta_0, \theta_1, \theta_2]^{\mathrm{T}} = [\sin(t), \sin(t), \sin(t), \sin(t), \cos(t)]^{\mathrm{T}}$. Output signal and errors are shown in figure 6,7,8,9. The control performance indicators are shown in the table 8.

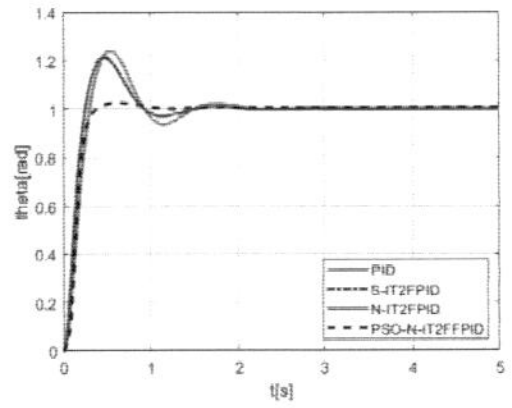

Figure 2: J_1 angle

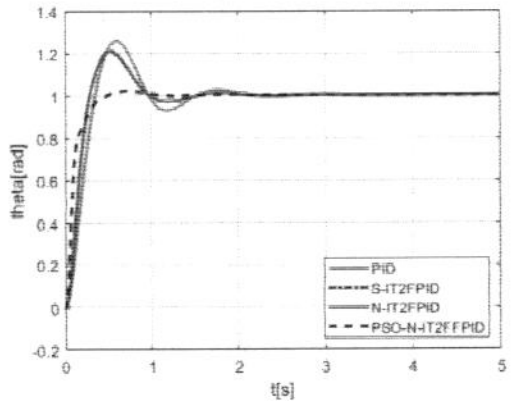

Figure 3: J_1 angle

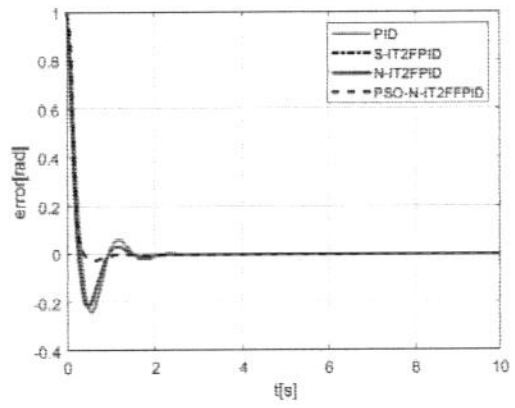

Figure 4: J_2 error

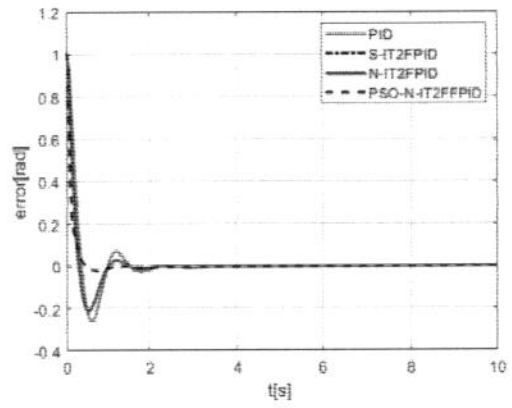

Figure 5: J_2 error

Table 7: Index of different control systems

control method	joint 1		joint 2	
	ITAE	IAE	ITAE	IAE
PID	13.219	28.771	16.454	32.005
S-IT2FPID	8.7065	23.274	10.511	24.909
N-IT2FPID	8.6299	23.029	10.422	24.631
PSO-N-IT2FPID	7.2198	18.229	6.2369	12.878

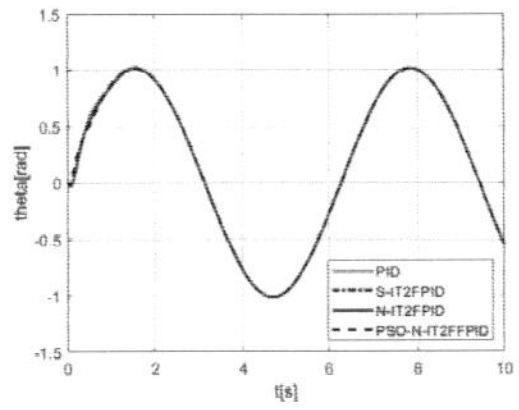

Figure 6: J_1 angle

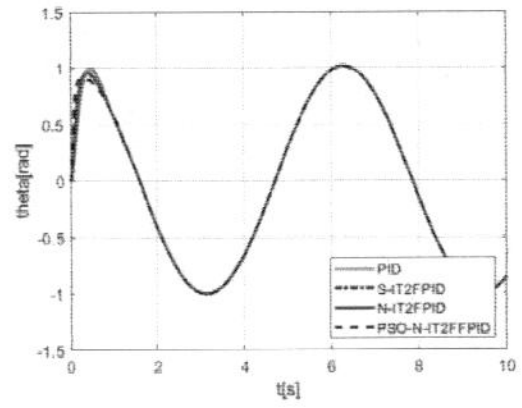

Figure 7: J_1 angle

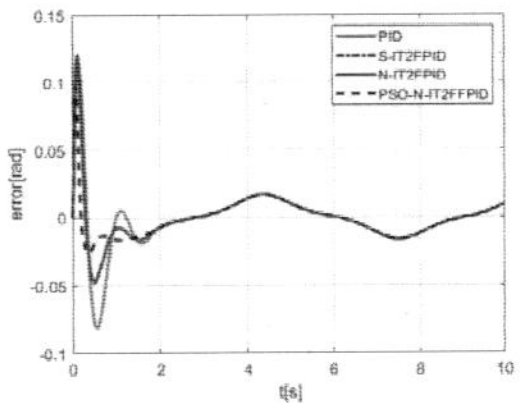

Figure 8: J_2 error

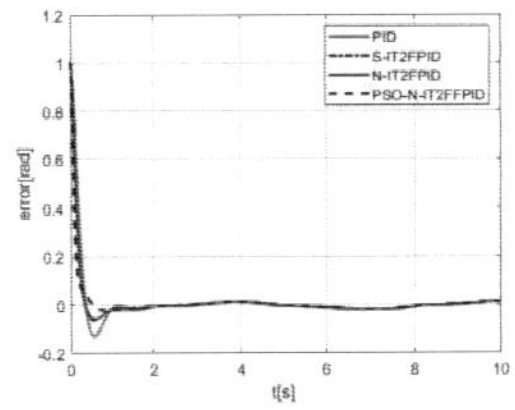

Figure 9: J_2 error

Table 8: Index of different control systems

control method	joint 1		joint 2	
	ITAE	IAE	ITAE	IAE
PID	6.9615	6.5009	9.7531	22.906
S-IT2FPID	6.2814	5.1911	8.1083	17.696
N-IT2FPID	6.2772	5.1641	8.0345	17.45
PSO-N-IT2FPID	5.9101	3.7235	7.0934	12.862

The same conclusion can be drawn by observing the data in table 8 when the input signal is sine. The performance of N-IT2FPID control system is better than the conventional IT2FPID control system. When the input signal is sine, PSO algorithm can also optimize the IT2FPID controller further.

6. Conclusions

Firstly, the dynamic model of flexible-link and flexible-joint space manipulator is given, and the N-IT2FPID controller is designed for the model, and the improved PSO method is also used. The numerical experiment show that for the tracking task of space manipulator, the performance of the proposed N-IT2FPID controller is better than that of singleton fuzzy method. The improved

PSO algorithm can markedly boost performance of the control system.

Acknowledgments
This work was supported in part by the Harbin Institute of Technology; in part by the National Natural Science Foundation of China under Grant 62188101, 62303135, and 62320106001; in part by the Heilongjiang Touyan Team Program. (Corresponding author: Yi Zeng)

References
[1] Hu Q, Zhang J. Maneuver and vibration control of flexible manipulators using variable-speed control moment gyros[J]. Acta Astronautica, 2015, 113: 105-119.
[2] Sharma R, Rana K P S, Kumar V. Performance analysis of fractional order fuzzy PID controllers applied to a robotic manipulator[J]. Expert systems with applications, 2014, 41(9): 4274-4289.
[3] Tong W, Zhao T, Duan Q, et al. Non-singleton interval type-2 fuzzy PID control for high precision electro-optical tracking system[J]. ISA transactions, 2022, 120: 258-270.
[4] Gaidhane P J, Nigam M J, Kumar A, et al. Design of interval type-2 fuzzy precompensated PID controller applied to two-DOF robotic manipulator with variable payload[J]. ISA transactions, 2019, 89: 169-185.
[5] Abdulameer H I, Mohamed M J. Fractional Order Fuzzy PID Controller Design for 2-Link Rigid Robot Manipulator[J]. International Journal of Intelligent Engineering & Systems, 2022, 15(3).
[6] Ardeshiri R R, Khooban M H, Noshadi A, et al. Robotic manipulator control based on an optimal fractional-order fuzzy PID approach: SiL real-time simulation[J]. Soft Computing, 2020, 24(5): 3849-3860.
[7] Liu Y, Jiang D, Yun J, et al. Self-tuning control of manipulator positioning based on fuzzy PID and PSO algorithm[J]. Frontiers in Bioengineering and Biotechnology, 2022, 9: 817723.
[8] Baz R, El Majdoub K, Giri F, et al. Fine-tuning quarter vehicle performance: PSO-optimized fuzzy PID controller for in-wheel BLDC motor systems[J]. IFAC-PapersOnLine, 2024, 58(13): 715-720.
[9] Fu X.Impact dynamics analysis, motion and vibration integrated control of flexible-base flexible-link and flexible-joint space robot capturing satellite[D].Fuzhou University,2022.
[10] Carvajal J, Chen G, Ogmen H. Fuzzy PID controller: Design, performance evaluation, and stability analysis[J]. Information sciences, 2000, 123(3-4): 249-270.
[11] Wu D. Approaches for reducing the computational cost of interval type-2 fuzzy logic systems: overview and comparisons[J]. IEEE Transactions on Fuzzy Systems, 2012, 21(1): 80-99.
[12] Tang H, Ma R, Chen Y, et al. Temperature regulation of hot vapor preservation treatment of litchi based on PSO-fuzzy PID[J]. Applied Sciences, 2023, 13(12): 6888.
[13] Marzaki M H, Tajjudin M, Rahiman M H F, et al. Performance of FOPI with error filter based on controllers performance criterion (ISE, IAE and ITAE)[C]//2015 10th Asian control conference (ASCC). IEEE, 2015: 1-6.

Design Method for Streamwise Area Distribution of Variable Cross-Section Curved Isolator

Bo Wang, Fei Qin, Zhengze Zhang*, Xianggeng Wei, Yajun Wang, Jinying Ye, Xing Sun, Zhen Tao

Internal Flow and Thermo-structure Laboratory, Northwestern Polytechnical University, Xi'an Shaanxi 710072, PR China

*Corresponding author's e-mail: zhangzhengze@nwpu.edu.cn

Abstract. The isolator, as a critical component in the rocket based combined cycle engine for matching the intake and combustion processes, plays a vital role in ensuring the efficient operation of the engine across a wide speed range. However, designing a curved variable-cross-section isolator with superior aerodynamic performance remains a significant challenge. This paper proposes a refined design method for the distribution of the streamwise cross-sectional area. For different isolator contours, the impact of this design method on the total pressure recovery capability and anti-backpressure capability is discussed. The results indicate that the refined design of the streamwise cross-sectional area can improve the total pressure recovery coefficient at the isolator outlet to some extent, while the anti-backpressure performance of the isolator can be enhanced by approximately 14.5%. The findings of this study can provide valuable insights for the wide speed range design of isolators.

1. Introduction

The rocket based combined cycle (RBCC) engine, as an ideal propulsion system for the next-generation reusable space-earth transportation vehicles, combines the high thrust-to-weight ratio of rocket engines with the high specific impulse of air-breathing ramjet engines. Its operational range spans subsonic, supersonic, and hypersonic regimes [1].

The wide-speed and wide-altitude characteristics of RBCC engines necessitate the provision of stable, high-quality incoming flow to the combustion chamber across a wide speed range. Additionally, considering the requirements of modular design and integrated airframe-propulsion systems, the inlet and isolator exit geometries are generally different [2–5], with a certain spatial offset [6–10]. Therefore, there is a pressing need to develop efficient and reliable contour design methods for wide speed range variable cross-section curved isolators.

The design methods for hypersonic engine internal flow paths primarily include the streamline tracing method and the geometric blending functions. The streamline tracing method has been applied to the design of hypersonic inlets [11–18]. By tracing the flow characteristics in the basic flow field, a specific streamline envelope surface is obtained, which serves as the internal flow boundary to guide subsequent geometric modeling. The geometric blending functions is typically used for transitioning between different inlet and outlet shapes [3–5, 19-21], without considering flow field characteristics, and is purely a mathematical method for surface transition.

Barger [19] proposed the geometric blending functions when designing transition geometries of variable cross-section forebody, achieving smooth transitions in variable cross-section geometric channels. Subsequently, Smart [20] conducted research on the design of REST supersonic three-dimensional inlets based on the geometric blending functions and streamline tracing method. Compared to traditional two-dimensional inlets, this three-dimensional inlet demonstrated higher total pressure recovery capabilities at both design and off-design points. The National University of Defense Technology [3–5, 21] conducted optimization design studies on variable cross-section curved isolators using both the streamline tracing method and the geometric blending functions. Both approaches employed the Multi-Island Genetic Algorithm (MIGA) and utilized the cross-sectional area variation law as an optimization objective. The maximum difference in the cross-sectional area distribution before and after optimization was approximately 4%. However, the optimization iteration cycles were lengthy, requiring 600–800 iterations to achieve results. To enhance design efficiency, there is a need to develop a direct design method for the cross-sectional area distribution.

To address the efficient design requirements for variable cross-section curved isolator contours, this paper proposes and validates a direct design method for the streamwise cross-sectional area distribution. The impact of this design method on the total pressure recovery capability and anti-backpressure capability of different isolator contours is discussed at a cruise trajectory point of a specific RBCC engine.

2. Calculation method and validation

2.1. Simulation method

The three-dimensional variable cross-section curved isolator contour is chosen as the research objective. The distributions of aerodynamic parameter are obtained by solving the three-dimensional compressible Reynolds-averaged Navier-Stokes (RANS) equations. The spatial discretization employs a second-order upwind scheme, while the temporal advancement utilizes an implicit time integration method. The SST-k-ω turbulence model is selected for the validation of the numerical method. During the grid independence verification for the numerical simulations, computational models with total grid counts of 30.8×10^4, 62.2×10^4, and 144×10^4 are used.

The numerical method in this study is validated using the experimental results from Kawatsu [22], as shown in Figure 1. The vertical axis represents the dimensionless wall static pressure, and the horizontal axis represents the dimensionless axial position. Detailed explanations of the parameters can be found in reference [22]. The experiment considered the influence of the isolator exit backpressure on the mainstream flow. It can be observed that the leading edge of the shock train in the experiment is approximately located at $x/H = 5$. The numerical results for the static pressure distribution along the flow path are in complete agreement with the experimental results in terms of trend, and the numerical curve accurately predicts the position of the shock train leading edge. The maximum discrepancy between the numerical results and the experimental data at the measurement points is around 16%. This numerical method is capable of predicting the isolator flow field with reasonable accuracy and is thus suitable for the subsequent numerical simulations in this study.

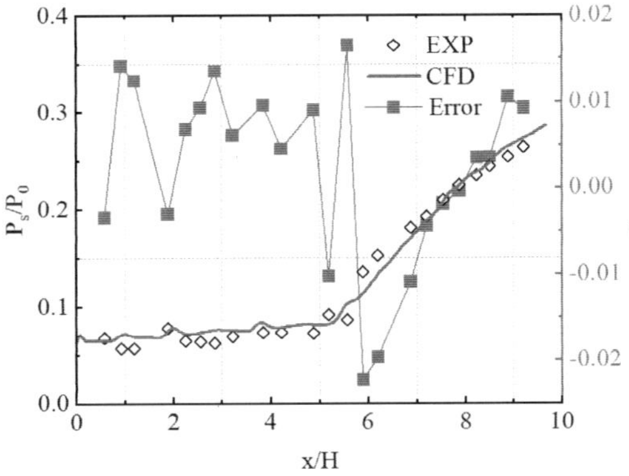

Figure 1. Comparation of numerical and experimental results

2.2. Mesh independency

The baseline isolator contour, as shown in Figure 2, is selected for the grid independence verification. The axial dimension of the isolator is 0.6 m, and the specific dimensions of the inlet and outlet are illustrated later in Figure 4. The inlet and outlet are defined as pressure boundary conditions, while the walls are treated as adiabatic no-slip solid boundaries. This study aims to investigate the influence of the isolator transition contour on the total pressure loss of the incoming flow. Therefore, the backpressure at the outlet is set to 100 Pa, consistent with the simulation setup in reference [3]. During grid generation, the thickness of the first layer of near-wall grids is maintained at 0.01 mm. The static pressure distribution along the centerline of the isolator's upper wall is extracted, with the specific location indicated by the red line in Figure 2.

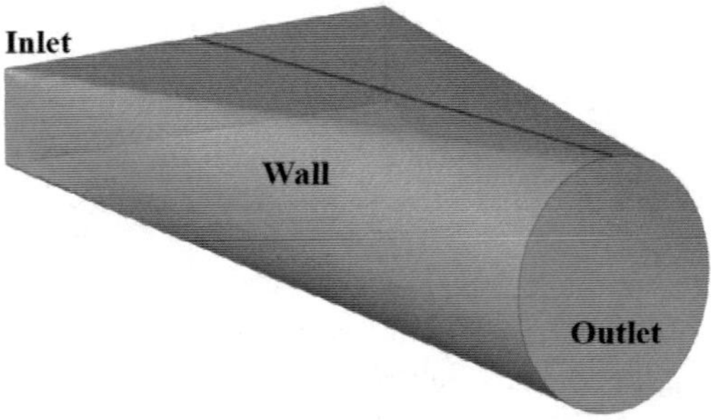

Figure 2. Baseline model of isolator

Figure 3 presents the corresponding results for different grid sizes, with the static pressure along the flow path normalized by the inlet static pressure. Qualitatively, it can be observed that the trend of the static pressure curves remains consistent across different grid sizes. However, as the grid count increases from 30.8×10^4 to 144×10^4, the static pressure level within the range of $x \leq 0.3$ gradually decreases, with the rate of decrease diminishing progressively. Compared to the results from other grid sizes, when the grid count increases from 62.2×10^4 to 144×10^4, a more than twofold increase in grid quantity, and the change in the static pressure curve along the flow path is minimal. Therefore, for subsequent numerical simulations, the grid generate strategy corresponding to the 62.2×10^4 grid count is adopted in this article.

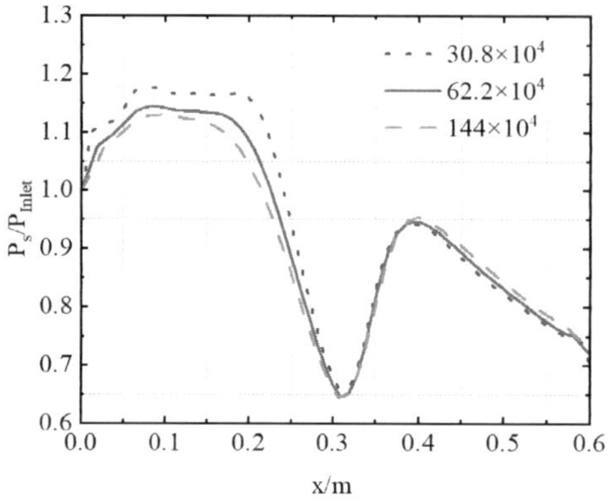

Figure 3. Static pressure along top wall

3. Contour design method

To address the requirement for a direct design method for the streamwise cross-sectional area distribution, this article proposes a direct control method based on the geometric blending functions introduced by Barger [19] and the variable cross-section curved isolator modelling method developed by Chen et al. [3].

Equation 1 represents the classical geometric blending formula [19], where the contour equation (*f(y)*) of each cross-section is influenced by the inlet and outlet contours $f_1(y)$ and $f_2(y)$ as well as the smoothing parameter α. The subscripts *ca* and *cc* denote the inlet and outlet contours, respectively. In this study, a scaling factor κ is introduced into the classical geometric blending formula. By proportionally scaling the discrete points of the cross-sectional contour using the scaling factor κ, the cross-sectional area is adaptively adjusted based on the bisection method to match the target cross-sectional area. This achieves direct control of the streamwise cross-sectional area distribution.

$$f(y) = \kappa[f_1(y)]^{1-E(x)}[f_2(y)]^{E(x)}$$

$$E(x) = \left(\frac{x - x_{ca}}{x_{cc} - x_{ca}}\right)^{\alpha} \quad \alpha > 0 \tag{1}$$

Figure 4 illustrates the inlet and outlet contours of the isolator used in this study. The outlet contour of isolator is circular with a diameter of R, while the isolator inlet contour is rectangular with dimensions of 0.37R×1.96R. The offset height between the centroid of the inlet contour and the outlet contour is 0.5R.

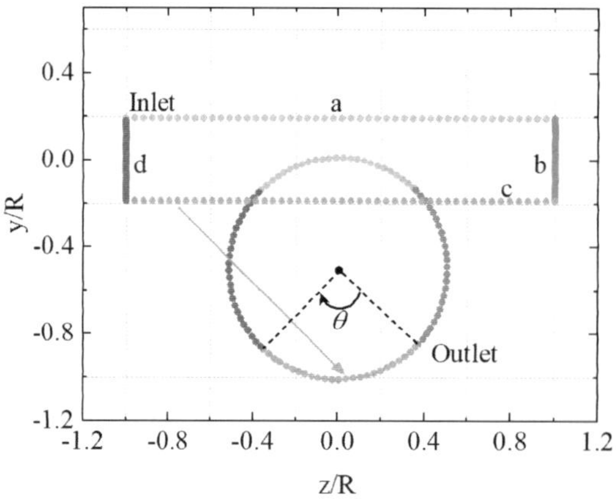

Figure 4. Discretized points of inlet and outlet

During the geometric blending process, the four vertices of the rectangular inlet contour are used as characteristic points, dividing the inlet contour into four segments: a, b, c and d. Correspondingly, the outlet contour is also divided into four matching segments. Considering the symmetry of the inlet and outlet contours, the sum of the angles corresponding to segments a and b of the outlet contour is fixed with 180°. In Figure 4, discrete points of the same color represent the corresponding inlet and outlet contours, where the dimensionless parameter θ denotes the central angle spanned by the outlet contour segment corresponding to the inlet segment c. The correspondence between the characteristic contours of the inlet and outlet is uniquely determined by the parameter θ.

When determining the centerline shape of the offset isolator, five control points are used for constraint, as shown in Figure 5. Points P_1 and P_5 are the centroids of the inlet and outlet contours, respectively. To ensure smooth airflow at the inlet and outlet, the centerline at both ends should be tangent to the axial direction. Therefore, the radial movement of points P_2 and P_4 is not considered, and their axial positions are determined using the dimensionless parameters x_2 and x_4. Point P_3 is always positioned at the midpoint of the line connecting P_1 and P_5 in the axial direction, and its radial position is controlled by the dimensionless parameter y_3.

For the geometric modelling of the variable cross-section curved isolator, fine control over the streamwise cross-sectional area distribution requires adjusting the spatial positions of the transition sections to varying degrees, as illustrated in Figure 5. In this study, Equation 2 is used to modify the coordinates of the discrete points on each cross-sectional contour, ensuring that the normal direction of each contour aligns with the tangent direction of the corresponding centerline (the red solid lines indicate the transformed cross-sectional positions). This results in a variable cross-section curved isolator with controlled streamwise cross-sectional area distribution. In Equation 2, the subscript p denotes the discrete points on the control contour of each cross-section, the subscript o represents the centroid position of each cross-section, the superscript * indicates the control points of each cross-section before coordinate transformation, and γ is the angle between the normal direction of each cross-section before coordinate transformation and the tangent direction of the corresponding centerline. The subscript i denotes the ith intermediate cross-section, and the subscript j denotes the jth discretized point of current cross-section.

$$
\begin{aligned}
R_{p\_j} &= \sqrt{\left(x^*_{p\_j} - x_{o\_i}\right)^2 + \left(y^*_{p\_j} - y_{o\_i}\right)^2} \\
x_{p\_j} &= R_{p\_j} \cdot cos\gamma_i \\
y_{p\_j} &= R_{p\_j} \cdot sin\gamma_i \\
z_{p\_j} &= z^*_{p\_j}
\end{aligned}
\tag{2}
$$

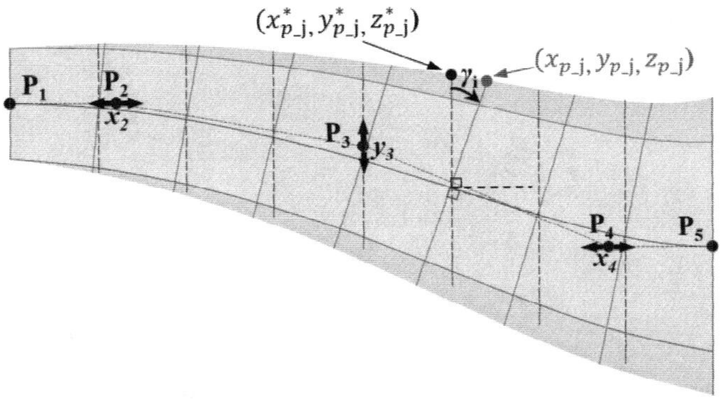

Figure 5. Front view of isolator

4. Results and discussion

4.1. Distribution of cross-sectional area

The design method of isolator contour proposed in this article includes a total of five dimensionless design parameters. Considering the rationality of the geometric contour, the ranges of these parameters are constrained as follows: α ranges from 1 to 3; θ ranges from 0.1 to 0.9; x_2 ranges from 0.1 to 0.5; x_4 ranges from 0.1 to 0.5; and y_1 ranges from 0.1 to 0.9. The total pressure recovery coefficient of the outlet flow quantitatively characterizes the total pressure recovery capability of the isolator. This coefficient, denoted as σ, is used to evaluate the aerodynamic performance of the isolator transition contours.

Table 1 presents five different combinations of design parameters used to design geometric models, validating the isolator contour design method proposed. In this article, the streamwise cross-sectional area distribution of the isolator follows a linear trend. Figure 6 shows the streamwise cross-sectional area distribution curves for different models before and after area correction, where the superscript * denotes the isolator contours after area correction. For different initial isolator models, the maximum deviation between the cross-sectional area distribution and the target distribution is approximately 30%. However, after correcting the streamwise cross-sectional area distribution, the agreement between the initial area distribution and the target distribution is significantly improved, with a maximum deviation of only about 1.1%. This demonstrates that the proposed isolator design method can accurately achieve direct design of the streamwise cross-sectional area distribution. In subsequent work, the streamwise cross-sectional area distribution of the isolator will consistently follow a linear trend.

Table 1. Combinations of parameters

Model	α	θ	x_2	x_4	y_3
$Model_1$	1	0.1	0.1	0.1	0.1
$Model_2$	1.5	0.3	0.2	0.2	0.3
$Model_3$	2	0.5	0.3	0.3	0.5
$Model_4$	2.5	0.7	0.4	0.4	0.7
$Model_5$	3	0.9	0.5	0.5	0.9

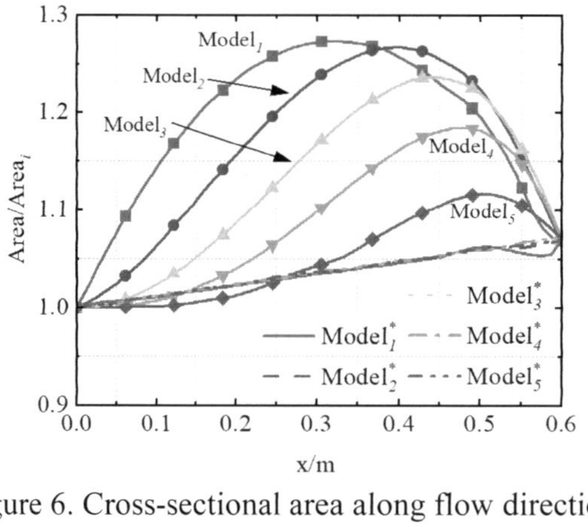

Figure 6. Cross-sectional area along flow direction

4.2. Total pressure recovery capability

A cruise trajectory point of a specific RBCC engine is selected to investigate the influence of the streamwise cross-sectional area distribution on the total pressure recovery capability of the isolator. As shown in Table 2, P_t and T_t represent the stagnation pressure and stagnation temperature at the isolator inlet, respectively, and Ma_∞ denotes the freestream Mach number at the trajectory point.

Table 2. Boundary condition

Ma_∞	Pt/kPa	Tt/K
6	2277	1665

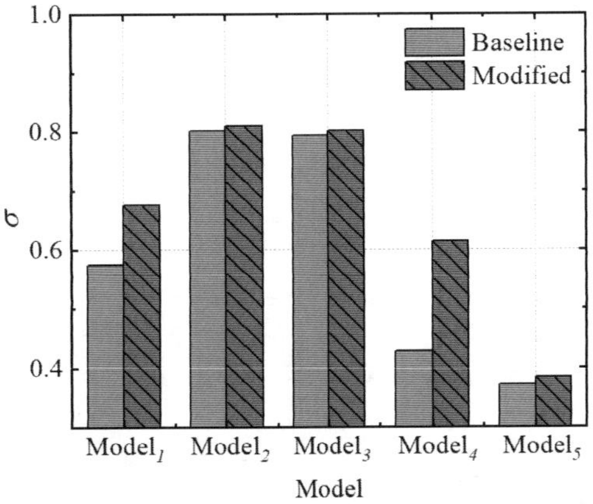

Figure 7. Total pressure recovery coefficients

Figure 7 presents the total pressure recovery coefficients at the outlet for different isolator contours. The modified models are obtained by applying the proposed method to control the streamwise cross-sectional area of the baseline models. It can be observed that the total pressure recovery coefficients vary significantly across different isolator contours. The maximum total pressure recovery coefficient of 0.80 corresponding to $Model_2$, and the minimum of 0.37 corresponding to $Model_5$, differ by approximately 0.43. After applying the direct design method proposed to regulate the streamwise cross-sectional area distribution of the isolator contours, the total pressure recovery coefficients at the isolator outlet show certain improvements compared to the initial contours.

After actively controlling the streamwise cross-sectional area distribution for $Model_2$ and $Model_5$, their total pressure recovery coefficients increase by 0.9% and 3.2%, respectively. However, the difference between the two total pressure recovery coefficients remains around 0.42. In Figure 6, the maximum difference between the area distribution curves of $Model_2$* and $Model_5$* is only 0.24%, indicating that their streamwise cross-sectional area distributions are essentially consistent. This suggests that the differences in the transition contours of the variable cross-section curved isolator have a significant impact on the total pressure recovery coefficient at the outlet, as shown in Figure 7, where the maximum difference can reach approximately 111.5%.

Figure 8 extracts the Mach number contours for multiple axial cross-sections of $Model_2$* and $Model_5$*. In Figure 8(a), the high Mach number regions occupy a relatively large proportion of each axial cross-section. After the incoming flow enters the isolator inlet, the proportion of high Mach number regions in cross-sections ② to ④ decreases slightly along the streamwise direction. However, the majority of the mainstream flow remains around Ma = 3. As the bottom wall of isolator expands, the local area near the lower wall in cross-section ⑤ increases rapidly, causing the local supersonic flow to expand and decelerate. When the flow exits the isolator outlet, the Mach number in the majority of the region remains around Ma = 3, with only a small low Mach number region appearing near the bottom wall.

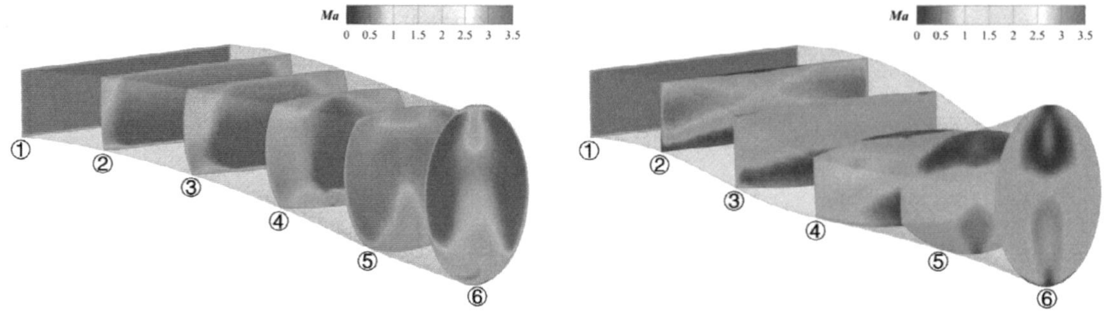

(a). $Model_2$* (b). $Model_5$*

Figure 8. *Ma* contours of models

Compared to Figure 8(a), the decrease of Mach number along the streamwise direction is more pronounced in Figure 8(b). Although the incoming flow in the isolator still exhibits a distinct high Mach number region around $Ma = 3$ in the upper part of cross-section ②, the rapid expansion of the isolator's bottom wall causes the supersonic flow in the lower part of cross-section ② to expand and decelerate to approximately $Ma = 0.5$. In subsequent cross-sections, the proportion of the high Mach number region around $Ma = 3$ further decreases. Additionally, after cross-section ④, the further expansion of the isolator's upper and lower walls results in noticeable low Mach number regions around $Ma = 0.5$ in both the upper and lower parts of cross-section ⑤, which develop toward the centerline along the streamwise direction. Compared to $Model_2$*, the overall Mach number at the outlet of $Model_5$* is significantly lower, indicating greater flow losses as the incoming flow passes through $Model_5$*. This observation is consistent with the conclusions drawn from Figure 7.

4.3. Anti-backpressure capability

To evaluate the impact of the design method proposed on another critical performance metric of the isolator, anti-backpressure capability. we focus on $Model_2$, which demonstrated the best total pressure recovery performance in the previous section. In assessing the anti-backpressure capability of the isolator, existing research often uses the limit backpressure as a quantitative measure. The limit backpressure is defined as the maximum outlet backpressure at which the shock train is not expelled from the isolator inlet.

In this study, the limit backpressure for the isolator contours is determined by gradually increasing the backpressure in increments of $P_b/P_i = 0.2$. As discussed in the previous section, the flow phenomena in variable cross-section curved isolators exhibit strong three-dimensional characteristics. To more accurately predict the influence of backpressure on the shock train, the massflow averaged static pressure P_s is extracted for each streamwise cross-section of $Model_2$ and $Model_2$*, as shown in Figure 9. Here, the horizontal axis represents the centroid coordinates of each streamwise cross-section, and the vertical axis represents the ratio of the local static pressure to the inlet static pressure P_s/P_i. When the dimensionless static pressure at $x = 0$ exceeds 1, it indicates that the shock train has been pushed out of the isolator inlet under the given backpressure, which may lead to inlet unstart.

In Figure 9, the trends in the streamwise static pressure distribution under different backpressure conditions indicate that the airflow undergoes continuous expansion and compression within the isolator, which may be related to the specific shock train structures induced by varying backpressures. In Figures 9(a) and 9(b), when the shock train has not yet been expelled from the inlet, the overall static pressure level of the airflow inside the isolator remains relatively low. The peaks observed in the middle section are likely due to shock compression, as seen in the cases of $Model_2$ at $P_b/P_i = 8.0$ and $Model_2$* at $P_b/P_i = 9.0$ and $P_b/P_i = 9.2$. However, once the shock train is pushed out of the inlet, the isolator is considered to have failed. At this point, the static pressure inside the isolator is entirely influenced by the backpressure, and the overall pressure level is significantly higher than in cases where the isolator has not failed, as observed in $Model_2$ at $P_b/P_i \geq 8.0$ and $Model_2$* at $P_b/P_i = 9.4$ and $P_b/P_i = 9.6$.

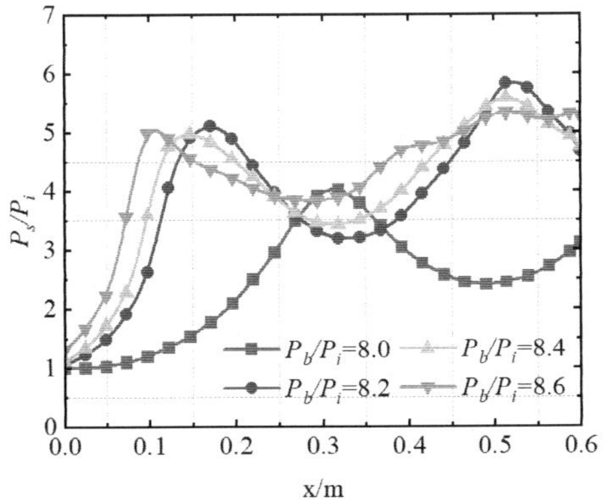

(a) $Model_2$

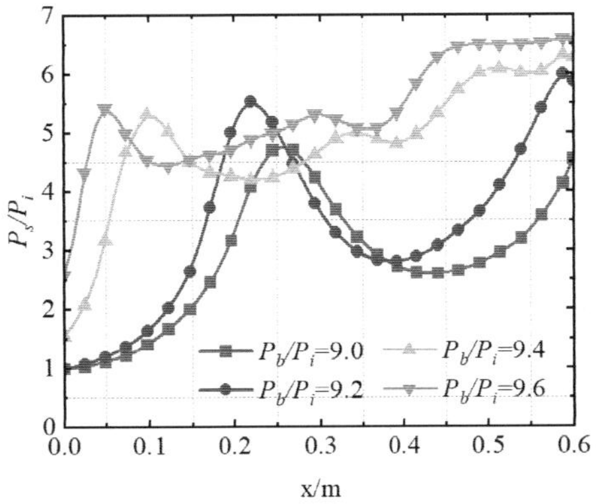

(b) $Model_2$*

Figure 9. Performance along flow direction

The limit backpressure ranges for different isolator contours are summarized in Table 3. After actively controlling the streamwise cross-sectional area distribution, the limit backpressure that the isolator can withstand increases by approximately 14.5%, reaching 1.2 times the inlet static pressure. This demonstrates that a linearly increasing streamwise cross-sectional area distribution can significantly enhance the anti-backpressure capability of variable cross-section curved isolators.

Table 3. Limit backpressure of $Model_2$

Model	Range of limit P_b/P_i
$Model_2$*	9.2-9.4
$Model_2$	8.0-8.2

5. Conclusions

This paper proposes a refined geometric blending method for designing the streamwise cross-sectional area distribution, develops a contour design method for variable cross-section curved isolators, and validates the effectiveness of this design method. The impact of this method on the total pressure recovery performance and anti-backpressure capability of different isolator contours is analyzed. The conclusions are as follows:

(1) By adaptively scaling the contours of each transition cross-section based on the traditional geometric blending theory, the maximum deviation between the initial and target cross-sectional areas at each streamwise location is only 1.1%. This achieves precise design over the streamwise cross-sectional area distribution in variable cross-section curved isolators.

(2) A linear streamwise cross-sectional area distribution can improve the total pressure recovery capability of the isolator to some extent. However, differences in the transition contours can cause the total pressure recovery coefficient at the isolator outlet to vary by more than 100%.

(3) The isolator design method proposed in this article can increase the limit backpressure by approximately 14.5%, significantly enhancing the isolator's anti-backpressure capability.

References

[1] Wang Y J, He G Q, Qin F, et.al. Research Progress of Rocket Based Combined Cycle Engines [J]. Journal of Astronautics, 2019, 40(10): 1125-1133.

[2] Zuo F Y, Mölder S. Hypersonic wavecatcher intakes and variable-geometry turbine based combined cycle engines[J]. Progress in Aerospace Sciences, 2019, 106: 108-144.

Journal of Physics: Conference Series **3056** (2025) 012002 doi:10.1088/1742-6596/3056/1/012002

[3] Chen J F, et al. Parameterization and Optimization for Variable Cross-Sections Bend Isolator. [J] Journal of Propulsion Technology 41.5: 1023.
[4] Meng Z, Fan X, Wang Y, et al. Parameterization and optimization for shape-transition curved isolator[J]. Acta Astronautica, 2018, 151: 563-571.
[5] Meng Z, Fan X, Wang Y, et al. Optimization design for shape-transition curved isolator with controllable cross-sectional area[J]. Acta Astronautica, 2018, 152: 335-341.
[6] Nie S, Ye J, Wei X, et al. Investigation on geometric throat matching characteristics of variable geometry rocket-based combined cycle combustor[J]. Acta Astronautica, 2023, 209: 104-116.
[7] XU Z, TAN J G, ZHANG D D. Ascent trajectory optimization for RBCC powered cruise vehicle [J] Journal of Rocket Propulsion, 2024, 50(2) : 49-56.
[8] Wang Z, Li Y H, Zhang M Z, et al. Simulation on flow field in isolator of RBCC engine with different configurations [J]. JOURNAL OF ROCKET PROPULSION, 2023, 49(6):82-89.
[9] Gong C, Bing C, Gu L, et al. The Airframe/Propulsion Integrated Design, Analysis and Ground Test for RBCC Launch Vehicle[C] 21st AIAA International Space Planes and Hypersonics Technologies Conference. 2017: 2318.
[10] Zhang T, Wang Z, Huang W, et al. The overall layout of rocket-based combined-cycle engines: a review[J]. Journal of Zhejiang University-SCIENCE A, 2019, 20(3): 163-183.
[11] Meng Z W. Parameterization and Optimization Design for Forebody and Inlet of Hypersonic Vehicle [D]. National University of Defense Technology, 2018.
[12] Tognelli A, Cakir B O, Saracoglu B H. Design and Investigation of Busemann Intakes for High-Speed Propulsion Systems[C] AIAA SCITECH 2023 Forum. 2023: 2353.
[13] Lim C, Jin S, Park G. Design and assessment of short-in-length shape transition hypersonic inlet with circular throat[J]. Journal of Mechanical Science and Technology, 2023, 37(11): 6047-6055.
[14] Clemence L K, White Z P, Kotler A R, et al. Variable Busemann Inlet Geometries for Hypersonic Vehicles[C] AIAA SCITECH 2023 Forum. 2023: 0122.
[15] Zuo F Y, Mölder S. Flow quality in an M-Busemann wavecatcher intake[J]. Aerospace Science and Technology, 2022, 121: 107376.
[16] McIntosh K A, Linton M, Rumpfkeil M P, et al. Experimental and Computational Study of Generic Busemann Inlets[C] AIAA SCITECH 2022 Forum. 2022: 0201.
[17] Xiong B, Fan X Q, Wang Y. Deficiencies of streamline tracing techniques for designing hypersonic inlets and potential solutions[J]. Science China Technological Sciences, 2020, 63(3): 488-495.
[18] Mölder S. The busemann air intake for hypersonic speeds[J]. Hypersonic Vehicles-Past, Present and Future Developments, 2019: 1-38.
[19] Barger R L. A procedure for designing forebodies with constraints on cross-section shape and axial area distribution[R]. 1981.
[20] Smart M K. Design of three-dimensional hypersonic inlets with rectangular-to-elliptical shape transition[J]. Journal of Propulsion and Power, 1999, 15(3): 408-416.
[21] Xiong B, Ferlauto M, Fan X. Optimal aerodynamic design of hypersonic inlets by using streamline-tracing techniques[J]. Adv. Aircr. Spacecr. Sci, 2020, 7(5): 441-458.
[22] Kawatsu K, Koike S, Kumasaka T, et al. Pseudo-shock wave produced by backpressure in straight and diverging rectangular ducts [C] AIAA/CIRA 13th International Space Planes and Hypersonics Systems and Technologies Conference. 2005: 3285.

Numerical Investigation of Combustion Regime and Heat Release Characteristics in RBCC Engine Under Different Dynamic Pressures

Zhen Tao, Fei Qin, Zhengze Zhang*, Yajun Wang, Jinying Ye, Bing Liu, Bo Wang
School of Astronautics, Northwestern Polytechnical University, Xi'an 710072, China
*Corresponding author's e-mail: zhangzhengze@nwpu.edu.cn

Abstract. The dynamic pressure is an important ballistic parameter that affects the operating characteristics of the air-breathing engine. For the purpose of extending the flight envelope, investigating the performance of RBCC engine under different dynamic pressures is of considerable value. The reactive flow is numerically investigated using an Improved Delayed Detached Eddy (IDDES) method. Three operating schemes with different dynamic pressures of 40, 60 and 80kPa are comparatively studied. Increasing dynamic pressure causes the supersonic flow region to be more significantly compressed, a phenomenon that results in a drop in the total flow field velocity, as demonstrated by the results. With dynamic pressure elevation, turbulent behaviors in the recirculation region are notably intensified. As dynamic pressure rises, the high adverse pressure gradient promotes the extension of the recirculation zone, thereby causing an expansion of the subsonic combustion area and a progressive increase in global combustion intensity. Additional analysis of the spatial distribution of combustion products demonstrates that an elevation in dynamic pressure can improve combustion efficiency within the combustor. According to statistical analysis, the subsonic heat release mode dominates the whole combustion heat release process, and the subsonic combustion intensity increases significantly with the increase of dynamic pressure. The heat release mode within the second stage combustor is more widely distributed than the first stage combustor.

1. Introduction

Recently, the increasing frequency of space exploration and development activities has placed higher demands on aerospace propulsion systems, requiring them to offer broad adaptability, cost-effectiveness, and high reliability[1, 2]. During the ascent from ground launch to orbital insertion, the propulsion system must cope with significant variations in velocity and altitude, necessitating strong adaptability to a wide range of dynamic pressures. Specifically, the system must achieve efficient fuel mixing and stable combustion across this broad dynamic pressure spectrum. The Rocket-Based Combined Cycle (RBCC) engine integrates the high thrust of rockets with the high specific impulse of ramjets, offering excellent wide-operating-envelope capability and making it one of the most promising propulsion systems for reusable space transportation. However, the large variations in inlet conditions pose substantial challenges to achieving rapid fuel mixing[3, 4], robust ignition[5], and stable combustion[6] within a single flow-path.

When the engine works in climbing trajectory, in order to obtain high thrust, the engine need to operate at high equivalence ratio injection strategy[7], which will lead to the increase of the overall combustion intensity of the combustor. There is obvious combustion oscillation in the combustor in

Journal of Physics: Conference Series **3056** (2025) 012003 doi:10.1088/1742-6596/3056/1/012003

strong combustion mode, and the change of engine dynamic pressure will lead to different levels of pressure changes in the combustor, and the corresponding combustion characteristics are also significantly different. Wang et al. [8] found that increasing the equivalence ratio will lead to obvious flame flashback in the combustor, and the combustion process will change from premixed combustion to premixed combustion and diffusion combustion, which is related to the increase of combustion intensity.

In order to study the supersonic combustion process more deeply, some scholars use different filter factors to classify the combustion modes, and obtain the combustion mode differences with different injection strategies, different flight speeds and different incoming components.

In this study, the flow field structure, combustion and heat release characteristics of a RBCC engine under different dynamic pressures are numerically investigated through IDDES method. The flow structures within the combustor are investigated firstly, on this basis, the flame propagation characteristics within the combustor are further analyzed. Furthermore, the heat release modes within the two-stage cavity are studied statistically.

2. Physical model and numerical method

2.1. Configuration and inflow conditions

Figure 1 gives the schematic diagram of the model combustor and the inlet flow parameters of the RBCC engine. The engine is mainly composed of isolator, combustor and nozzle. This study employs the isolator inlet height H to nondimensionalize the engine's structural parameters. The isolator includes two sections: a constant-area segment measuring 6.5H in length and a slightly expanding segment with a length of 10H. The combustor and nozzle have dimensions of 15H and 8H, respectively. Because of the symmetry of engine, in order to improve computational efficiency and save computational resources, half of computational domain are used for numerical simulation in this study.

The simulations study the Mach 6 flight conditions, and the flight pressure was 40 kPa, 60 kPa and 80 kPa respectively, and the total temperature was about 1650 K. The composition of the air inlet in the calculation basin is consistent with that of the alcohol/air heater, and the composition contents are $Y_{O_2} = 0.22$, $Y_{CO_2} = 0.12$,, $Y_{N_2} = 0.584$,and $Y_{H_2O} = 0.076$. With the change of dynamic pressure, the relevant parameters will be adjusted accordingly, the detailed engine inlet parameters are shown in Figure 1. This research utilized a two-stage strut injection strategy with fixed equivalence ratios of 0.6 (first stage) and 0.4 (second stage), aimed at independently investigating the impact of dynamic pressure on combustion.

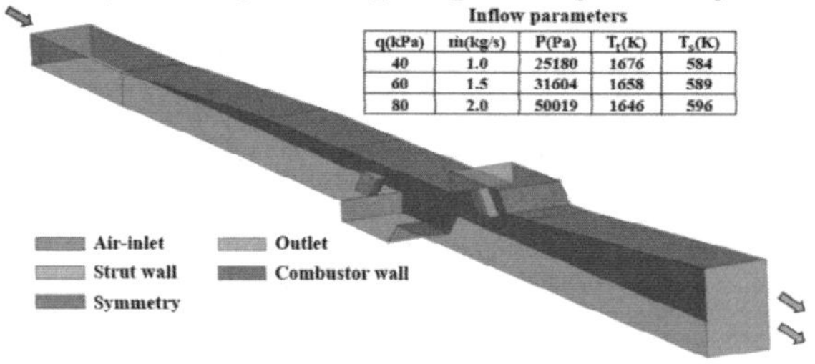

q(kPa)	$\dot{m}$(kg/s)	P(Pa)	T_t(K)	T_s(K)
40	1.0	25180	1676	584
60	1.5	31604	1658	589
80	2.0	50019	1646	596

Figure 1. Computational domain (half-model) and inflow parameters

2.2. Numerical method

The Improved Delayed Detached Eddy Simulation (IDDES)[9] method is a branch of hybrid RANS/LES method, the bulk of the flow is solved using the LES model, while the near-wall boundary layer region is resolved using RANS. IDDES fundamentally modifies the dissipation term in the k equation using the turbulent length scale, which can be formulated as follows:

$$\begin{cases} \dfrac{\partial(\rho k)}{\partial x} + \nabla \cdot (\rho U k) = \nabla \cdot [(\mu + \sigma_{kK} \mu_t) \nabla k] + P_k - \dfrac{\rho \sqrt{k^3}}{l_{IDDES}} \\ \dfrac{\partial(\rho \omega)}{\partial x} + \nabla \cdot (\rho U \omega) = \nabla \cdot [(\mu + \sigma_{\omega} \mu_t) \nabla \omega] + 2(1 - F_1) \rho \sigma_{\omega 2} \dfrac{\nabla k \cdot \nabla \omega}{\omega} + \alpha \dfrac{\rho}{\mu_t} P_K - \beta \rho \omega^2 \end{cases} \quad (1)$$

Where μ, μ_t, k, t, ρ, ω, U, F_l, P_K and l_{IDDES} represent dynamic viscosity, turbulent viscosity, turbulent kinetic energy, time, density, specific rate of dissipation, velocity, blending function, production term, and IDDES length scale, respectively. σ_k, σ_ω, $\sigma_{\omega 2}$, α and β are the model constants. Several researchers have validated the accuracy of the IDDES method for simulating supersonic combustion processes.

The turbulent combustion process is solved by the PaSR combustion model, which are conducted by OpenFOAM with second order accuracy. The PaSR combustion model captures the interaction between large-scale vortices and sub-grid-scale turbulence, while facilitating the coupling of turbulence and flame dynamics and is capable of simulating complex chemical kinetics. Its effectiveness has been validated through multiple studies. The turbulent combustion of kerosene/air is modeled using a simplified multi-step overall chemical reaction mechanism, with kerosene represented by the alternative molecular formula $C_{10}H_{22}$, comprising 11 species and 10 reaction steps.

To evaluate grid independence, 4 million, 6 million, and 9 million cells were employed to compute static pressures along the combustor wall centerline. Structural grid details for the combustor are illustrated in Figure 2. With the turbulent effects of the fluid and the capture of vortex flow patterns being considered, grid refinement is applied within the main flow region of the combustor, and the cavity interior.

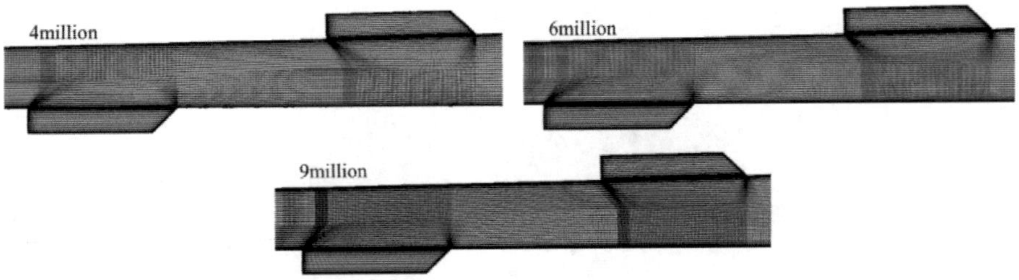

Figure 2. Computational grids with varying levels of refinement

As shown in Figure 3 (a), the wall pressure calculated under different grid resolutions exhibits a similar trend. However, the coarse grid does not accurately resolve the shock wave intensity and the pressure change trend in the combustor, whereas the results from the medium grid and the fine grid are highly consistent. Figure 3 (b) present the comparison between experiment and simulation results. From a quantitative perspective, wall pressure predictions using the IDDES method show strong correspondence with experimental observations, implying that the proposed computational framework is capable of precisely reconstructing the combustion flow field. In the following study, the above calculation method will continue to be used to study the flow, combustion and heat release characteristics of the combustor under different dynamic pressures.

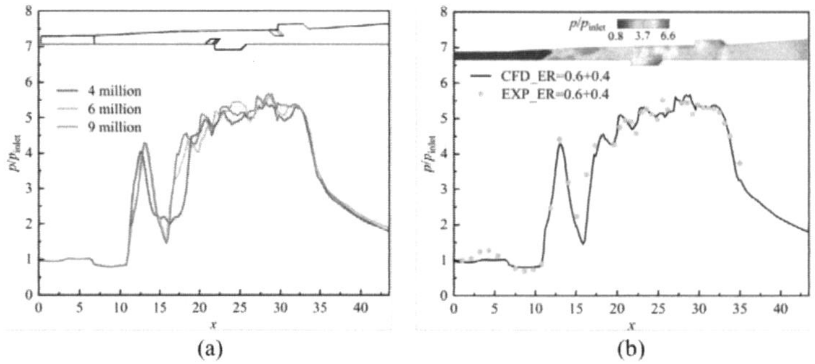

Figure 3. (a) Grid independence verification for grids with different levels of refinement; (b) Comparison between experiment and calculated wall pressure

3. Results and discussion

3.1. Flow field structures

Fig. 4 shows the instantaneous Mach number contour on the engine's middle plane under different dynamic pressures. A long low-speed separation zone forms near the lower wall. As the dynamic pressure increases, the leading edge of the separation zone gradually moves upstream and extends. The extension of the separation zone further compresses the supersonic flow region, reduces the overall flow field velocity, and significantly enlarges the subsonic flow region. As the dynamic pressure increases, the high pressure generated by combustion leads to a more severe adverse pressure gradient, which in turn causes the extension of the separation zone. Consequently, the combustion process becomes one of the key factors influencing the flow conditions within the combustor.

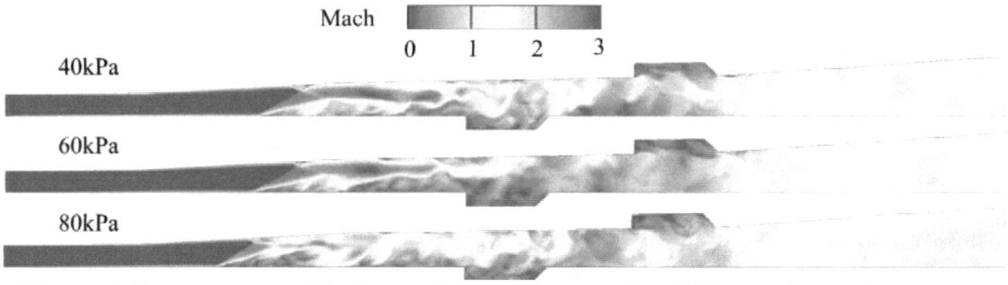

Figure 4. Instantaneous Mach number contours under different dynamic pressure

Figure 5 illustrates the instantaneous recirculation zone structures inside the combustor under different dynamic pressures, characterized by $U = -10m/s$ iso-surfaces. The iso-surfaces are colored by temperature to highlight the coupling between flow structures and combustion within the recirculation zones. The background shows the density gradient of the symmetry plane. It can be observed that as the dynamic pressure increases, the volume of the recirculation zone within the first-stage combustor significantly expands, and the turbulent flow characteristics within it become more significant. Under turbulent and flame oscillatory effects, the temperature field within the recirculation zone demonstrates marked spatial variations across distinct dynamic pressure settings. The turbulent eddies in this zone offer conducive flow patterns for flame spread, inducing a progressive upstream migration of high-temperature regions toward the central flow mainstream. However, for the second-stage combustor, although the impact of increased dynamic pressure on the flow field structure remains significant, the overall recirculation characteristics have been notably weakened. Within the second-stage cavity, notable recirculation features and elevated-temperature areas are observed, which implies a shift in the second-stage combustor's flame stabilization mode toward cavity-dominated combustion.

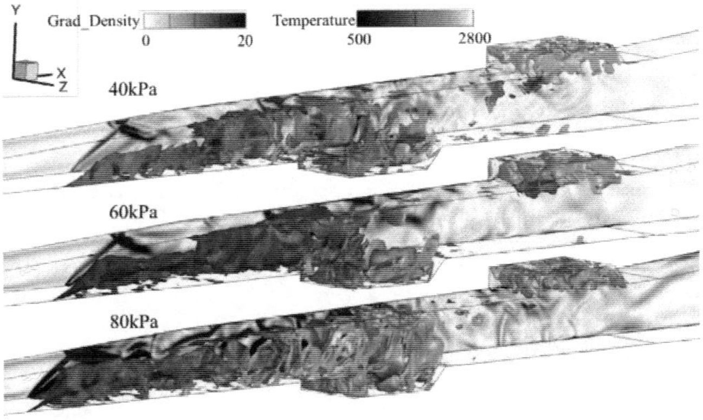

Figure 5. The structure of instantaneous recirculation zone in combustor ($U=-10m/s$ iso-surface) is colored by temperature and the background is density gradient.

To facilitate a quantitative comparison of flow characteristic variations induced by changes in dynamic pressure, the sectionally mass-averaged Mach number and pressure is used for analysis, with the calculation formula given as follows:

$$\begin{cases} \overline{Ma}_x = \dfrac{\int \rho \cdot U \cdot Ma \cdot dA}{\int \rho \cdot U \cdot dA} \\ \bar{p} = \dfrac{\int \rho \cdot U \cdot p \cdot dA}{\int \rho \cdot U \cdot dA} \end{cases} \tag{2}$$

The parameters in the equation (2) are time-averaged results. As shown in Figure 6 (a), differences in the Mach number distribution under varying dynamic pressures appear at the center of the isolator expansion section. As the dynamic pressure increases, it leads to the expansion of the separation zone, which in turn compresses the supersonic flow region, causing the Mach number in the flow field to decrease rapidly. The Mach numbers for the three cases follow the relationship $\overline{Ma}_{40kPa} > \overline{Ma}_{60kPa} > \overline{Ma}_{80kPa}$ once the flow stream enters the combustor. This also leads to the gradual expansion of the subsonic combustion zone as the dynamic pressure increases. Figure 6 (b) illustrates that the statistic pressure inside the combustor significantly rises. This is because the increase dynamic pressure directly enhances the overall pressure within the combustor, intensifies the combustion process, and consequently further increases the pressure level.

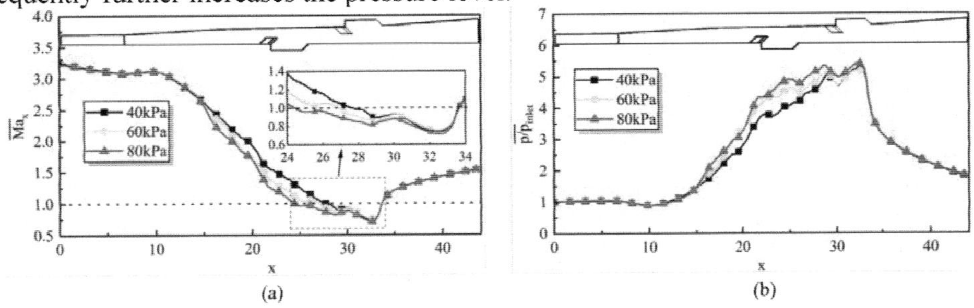

Figure 6. Time-averaged (a) Mach number ; (b) pressure distribution under different dynamic pressure

The fuel-air mixing characteristics are analyzed by mixture fraction distributions at the symmetry plane, as well as on some representative planes. The mixture fraction is defined as follows:

$$Z = \frac{sY_F - Y_O + Y_O^0}{sY_F^0 + Y_O^0} \tag{3}$$

Here, s represents the stoichiometric mass ratio, Y_F and Y_O denote the mass fractions of fuel and oxidizer, respectively. As depicted in Figure 7, under different dynamic pressures, the flow fields within the first-stage cavity exhibit a fuel-rich state. Particularly at 40 kPa and 60 kPa, the phenomenon of fuel propagating forward through the separation zone can be observed. As the dynamic pressure increases, the fuel becomes more concentrated within the cavity. In the second-stage strut/cavity region, a strip-shaped fuel-rich zone forms behind the strut, indicating that the coupling of the strut and cavity in the second-stage combustor is crucial for flame stabilization.

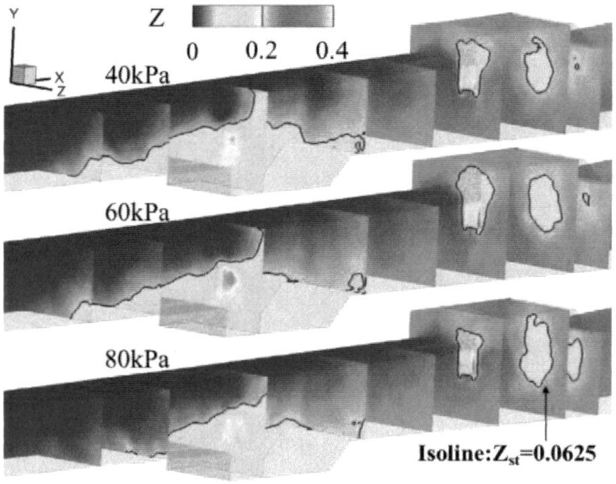

Figure 7. Time-averaged mixture fraction distribution under different dynamic pressures

3.2. *Combustion and heat release characteristics*

The differences in supersonic combustion characteristics under varying dynamic pressures can be directly characterized by the static temperature within the combustor. Figure 8 illustrates the instantaneous temperature distribution inside the combustor and the mean temperature distribution along the flow path. Overall, increasing dynamic pressure significantly enhances combustion intensity. At high dynamic pressures, the static temperature distribution is markedly higher compared to low dynamic pressure conditions. Locally, the phenomenon of flame propagation is more evident under high dynamic pressure, with near-wall recirculation zones facilitating flame propagation. Furthermore, the substantial rise in combustor pressure shifts the position of the pre-combustion shock train forward and increases its strength as dynamic pressure rises.

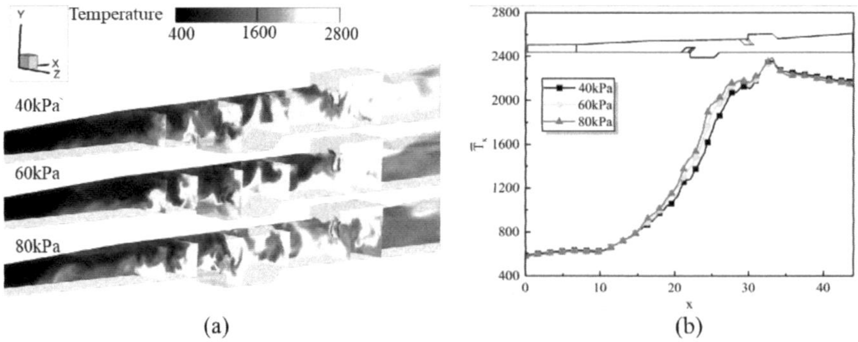

Figure 8. (a) Contours of instantaneous temperature distribution; (b) the one-dimensional distributions of time-average temperature

In this paper, $C_{10}H_{22}$ is used as the fuel. It first decomposes into the intermediate component CO, and then CO reacts to form the final component CO_2. Therefore, to evaluate the combustion process quantitatively, further comparative analysis of CO and CO_2 mass fraction distributions under different operational conditions is performed. The mass-weighted average technique for radical species is evaluated across the combustor's entire domain using

$$\bar{Y}_i(x) = \frac{\int Y_i \cdot \rho \cdot U \cdot dA}{\int \rho \cdot U \cdot dA} \tag{4}$$

As shown in Figure 9 (a), the CO mass fraction, the CO mass fraction experiences two cycles of initial increase followed by a decrease, which is related to the fuel injection and subsequent combustion consumption. For different operating conditions, as dynamic pressure increases, the overall distribution of CO shows a decreasing trend. This is due to the rise in combustor pressure, which accelerates the overall chemical reaction rate, speeding up both the consumption of CO, resulting in a decrease in the accumulated CO mass fraction within the combustor. As shown in Figure 9 (b), the accelerated combustion rate leads to more CO participating in reactions to form the final product CO_2. Consequently, an increase in dynamic pressure leads to higher CO_2 mass fraction.

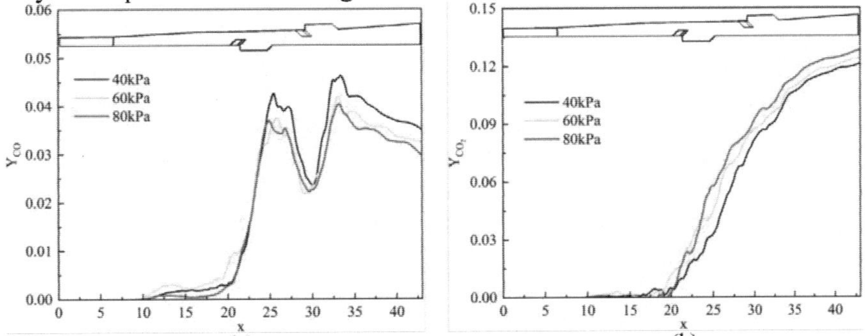

(a) (b)

Figure 9. Distributions of mass-weighted averaged CO and CO2 mass fraction.

To further investigate the statistical characteristics of heat release, a mapping relationship between subsonic/supersonic combustion and the heat release rate was established by a filter function, with the Mach number serves as the filter factor. The formulas are as follows:

$$\begin{cases} HR_{SUB} = \int \frac{(1 - sign(Ma-1))}{2} HRR \cdot dV \\ HR_{SUP} = \int \frac{(1 + sign(Ma-1))}{2} HRR \cdot dV \end{cases} \tag{5}$$

$$\begin{cases} Volume_{SUB} = \int \frac{(1 - sign(Ma-1))}{2} \cdot dV \\ Volume_{SUP} = \int \frac{(1 + sign(Ma-1))}{2} \cdot dV \end{cases} \tag{6}$$

Figure 10 illustrates the heat release distribution under different dynamic pressure conditions, including total heat release and the heat release rates corresponding to subsonic combustion and supersonic combustion. Overall, with increasing dynamic pressure, the heat release in both first and second stage combustor increases significantly, highlighting the dual-peak heat release distribution characteristics. In the first-stage combustor, both subsonic and supersonic combustion heat release modes are present. In contrast, the second-stage combustor predominantly exhibits the subsonic combustion mode. Following the second-stage cavity, however, the absence of effective flame stabilization devices causes the heat release mode to transition entirely to a supersonic combustion mode.

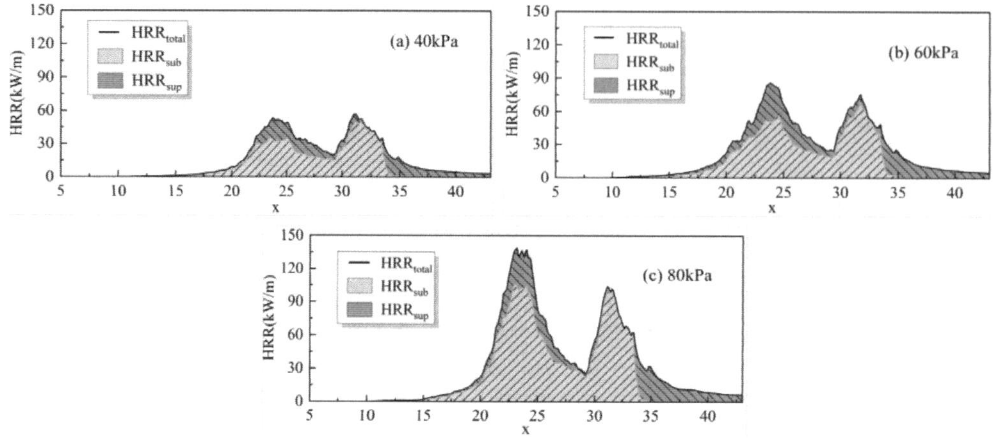

Figure 10. Heat release distribution along the streamwise direction under different dynamic pressures.

Figure 11 (a) and (b) display the quantity and the zone volume for subsonic and supersonic combustion heat release rate under different dynamic pressures. It is revealed by statistical analysis that subsonic combustion represents over 64% of the heat release quantity and 58% of the heat release volume under all dynamic pressure cases, implying that the subsonic combustion process is the primary driver of the overall heat release process. Further analysis shows that the quantity of heat release from subsonic combustion increases during the linear increase of the dynamic pressure, however, its heat release volume and volume proportion decrease continuously. This observation implies that heat release intensity in the subsonic combustion area is significantly augmented with increasing dynamic pressure.

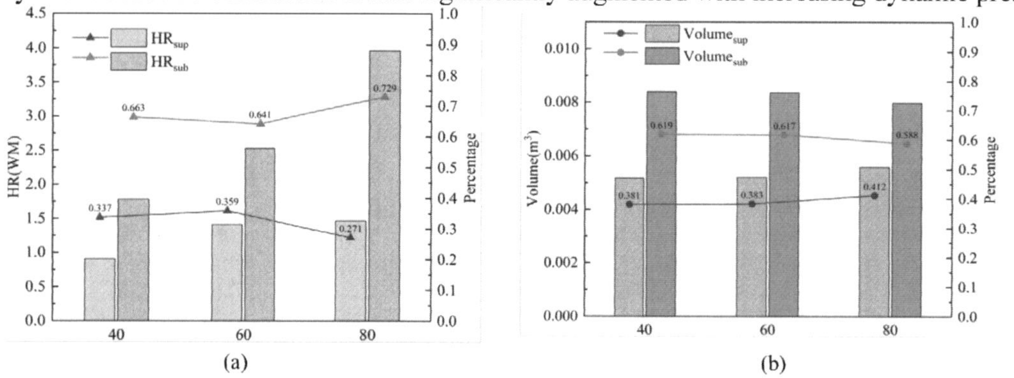

Figure 11. Heat release characteristics: (a) heat release quantity; (b) heat release volume.

To provide a detailed analysis of heat release statistical patterns, Figure 12 and Figure 13 present corresponding bivariate kernel density plots depicting the Mach number and heat release rate logarithm in the first-stage and second-stage combustors, respectively. As shown in Figure 12, the heat release in the first-stage combustor is concentrated within the subsonic range of $0.2<Ma<0.5$, with the heat release core located at $Ma=0.3$ that remains largely unchanged as dynamic pressure increases. This indicates that the first-stage combustor predominantly operates in a subsonic combustion mode. However, as the operating dynamic pressure increases, the high heat release zone gradually expands, suggesting that higher-pressure regions contribute more significantly to the total heat release. The heat release density distribution curve on the right side of the figure shows that an increase in dynamic pressure shifts the entire heat release distribution upward, indicating that higher dynamic pressure directly enhances the overall heat release intensity.

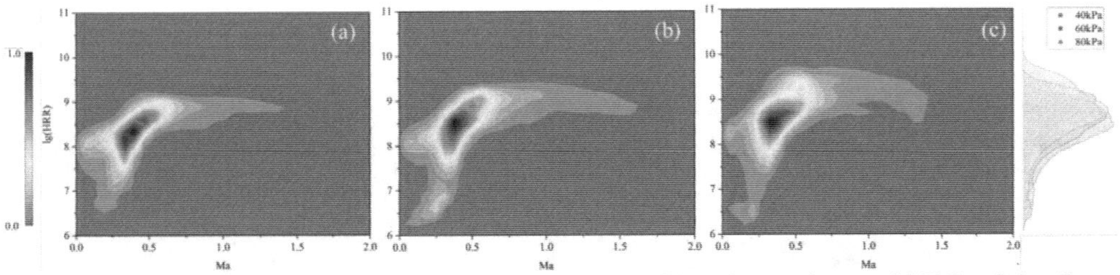

Figure 12. Bivariate kernel density estimation diagrams of Mach number and HRR of the first-stage cavity in (a) 40kPa, (b) 60kPa and (c) 80kPa; the probability distribution profile of HRR is on the right.

Figure 13 shows the combustion state of the second-stage combustor. It can be observed that compared to the first-stage combustor, both the Mach number distribution and the heat release rate have significantly expanded, with the heat release distribution becoming more uniform. Although the overall heat release rate has increased, it remains within the subsonic range (Ma < 1), indicating that the second-stage combustor operates primarily in a subsonic combustion mode. Similar to the first-stage combustor, the heat release distribution in the second stage follows the same trend. However, the heat release rate distribution in the second stage is wider, which suggests that even though the overall heat release distribution is similar, the combustion process inside the second-stage combustor is more intricate.

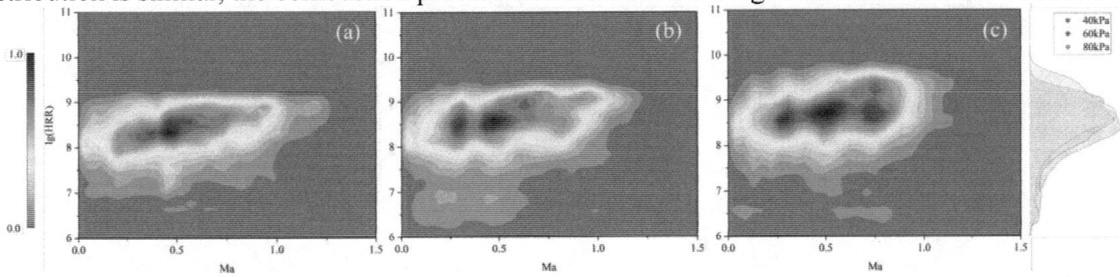

Figure 13. Bivariate kernel density estimation diagrams of Mach number and HRR of the second-stage cavity in (a) 40kPa, (b) 60kPa and (c) 80kPa; the probability distribution profile of HRR is on the right.

4. Conclusions

In this paper, an IDDES method is conducted to investigate the flow, combustion and heat release characteristics of the RBCC engine under different dynamic pressures. Through the numerical approach, the combustion process can be accurately investigated and deeply analyzed, which contributes to a more profound understanding of combustion behaviors in RBCC engines under varying dynamic pressures.

The separation zone expands as dynamic pressure increases, compressing the supersonic flow region, which in turn leads to a decrease in the overall flow velocity. Increased dynamic pressure also leads to an enlargement of the recirculation zone and enhances the coupling effect between the recirculation zone and combustion. For the second-stage combustor, the recirculation zone exists only within the cavity, so the second-stage combustor operates in a cavity stabilization mode. Analysis of time-averaged temperature, CO and CO_2 mass fraction distribution along the flow path indicates that increasing dynamic pressure significantly enhances overall combustion efficiency. Statistical analysis reveals that an increase in dynamic pressure significantly enhances heat release intensity, especially within the cavity region. In multi-regime combustion process, subsonic heat release mode occupies more than 58% of the volume and contributes over 64% of the heat release, thereby playing a key role in the combustion process. The kernel density plot reveals that as dynamic pressure increases, the overall heat release intensity also rises. Both the first and second combustors operate under subsonic combustion mode. However, the heat release density in the first-stage combustor is more concentrated, whereas in the second-stage combustor, the distribution is more dispersed.

Acknowledgments
This research work is financially supported by the National Natural Science Foundation of China under Grant No.52202482. The authors would like to thank NSFC for the sponsorship.

References

[1] Zeng Jia, Huang Hui, Zhu Pingping, et al. Research Progress and Key Technologies of Rocket-Based Combined Power[J]. Aerospace Overall Technology, 2022, 6(3): 9.
[2] Li, F., Liu, M. J., Sun, M. B., et al. (2025). Sensitive factors of ethylene combustion heat release under different combustion modes in scramjet engines. Acta Aeronautica et Astronautica Sinica, 46(4), 99-111.
[3] Wu, L. H., Shi, X. J., Li, Y., et al. (2024). Research progress on optical measurement techniques for scramjet combustors. Metrology & Measurement Technology, 44(3), 57-71.
[4] Feng, H., Jin, T., Wu, K. (2025). Numerical study on turbulent combustion of high-mach scramjet engines considering thermodynamic non-equilibrium effects. Journal of Experiments in Fluid Mechanics, 39(1), 30-37.
[5] Xu, D. Y., Tang, T., Yu, J. F., et al. (2024). Study on the influence of boundary layer thickness on supersonic flow combustion. Journal of Engineering Thermophysics, 45(2), 597-604.
[6] Wang, Z. C., Hu, B., Wang, Z. H., et al. (2024). Prediction model of transverse spray trajectory and dynamic characteristics analysis of supersonic inflow in ramjet combustor. Journal of Propulsion Technology, 45(5), 137-149.
[7] Yu, J. F., Tang, T., Yan, B., et al. (2024). Analysis of flow and combustion characteristics of circular scramjet combustor at Mach 6 flight conditions. Acta Aeronautica et Astronautica Sinica, 45(14), 221-234.
[8] Wang, Z. C. (2023). Study on flameout characteristics of wide-range engine ram combustor based on characteristic time scale. Thesis, Inner Mongolia University of Technology.
[9] D. Cao, H.E. Brod, N. Yokev, D. Michaels, Vortex dynamics in different combustion regions of a cavity-based scramjet, Proceedings of the Combustion Institute 39(3) (2023) 3147-3156.

Journal of Physics: Conference Series **3056** (2025) 012004 doi:10.1088/1742-6596/3056/1/012004

Application of Deep Learning in Flow and Heat Transfer Simulation of Supercritical Hydrocarbon Fuels

Li Yeyi[1,2], Zhang Minguo[1], Chen Rui[1,2], Ding Ziyan[1,2], Jiang Rongqing[1], Jing Tingting[1], Qin Fei[1], Liu Yinlong[1], Sun Xing[1]*

[1] National Key Laboratory of Solid Rocket Propulsion, School of Astronautics, Northwestern Polytechnical University, Xi'an, Shaanxi 710072, China

[2] Honors College, Northwestern Polytechnical University, Xi'an, Shaanxi 710072, China

*Corresponding author's e-mail: sunxing@nwpu.edu.cn

Abstract. With the rapid development of high-speed airbreathing engine technology, regenerative cooling has become the main solution to address the thermal protection issues of combustor wall. However, the complex physicochemical phenomenon during flow and heat transfer of supercritical hydrocarbon fuels, along with the high computational cost of traditional numerical simulation methods, severely limit the engineering application of this technology. In recent years, deep learning (DL), with its powerful nonlinear modeling capabilities and efficient computational advantages, has provided new insights for the study of supercritical flow and heat transfer problems. This paper reviews the current application of deep learning in the solution of flow equations, prediction of mixture thermophysical properties, acceleration of chemical reactions, and analysis of macroscopic heat transfer characteristics. Studies have shown that methods based on models such as physics-informed neural networks (PINN) and convolutional neural networks (CNN) can significantly improve computational efficiency, reduce wall temperature prediction errors, and effectively capture the multiscale features of supercritical fluids. However, existing methods still face challenges such as high data dependency and insufficient model generalization capabilities.

1. Introduction

With the rapid development of hypersonic vehicle technology, the thermal management issues of air-breathing engines have become increasingly prominent. At hypersonic speeds (Mach number exceeding 10), conventional passive cooling methods become insufficient to satisfy the thermal management demands of engine components exposed to extreme temperatures. Regenerative cooling technology, whch uses onboard hydrocarbon fuels as coolants, has become the main cooling method due to its excellent efficiency and cost-effectiveness [1-4].

However, the application of regenerative cooling technology into practical engine design process still faces numerous challenges. Firstly, the flow and heat transfer characteristics of hydrocarbon fuels under supercritical pressure are extremely complex, especially near the pseudocritical temperature, where the thermophysical properties of the fuel, such as density, viscosity, thermal conductivity, and specific heat capacity, undergo significant changes[5], leading to phenomena of enhanced or deteriorated heat transfer properties[6,7]. Secondly, traditional numerical simulation methods, such as computational fluid dynamics, require large computational resources and long processing times [8], making it difficult to meet the fast iteration requirements in engineering applications[9]. Especially in large eddy simulation (LES) and direct numerical simulation (DNS), the consumption of computational resources is unacceptable, which limits their application in high-precision simulations.

In recent years, the application of deep learning in fluid dynamics and heat transfer has shown promising prospects. Deep learning not only accelerates the solution of flow equations, but also enables rapid prediction of mixture thermophysical properties, acceleration of chemical reaction solutions [10], and prediction of macroscopic heat transfer characteristics such as convective heat transfer coefficients. This paper will investigate and review these four aspects, exploring the potential applications of deep learning in regenerative cooling technology and its future development directions.

2. Current Application of DL in Flow and Heat Transfer Prediction of Supercritical Hydrocarbon Fuels

2.1. Solving flow equations

Solving the Navier-Stokes equations often requires massive computational resources, especially when dealing with complex geometries, turbulence, and multi-physics field coupling. Traditional numerical methods often involve large computational costs, limited accuracy, and long solution times. Therefore, accelerating the solution of these equations while maintaining high accuracy has become an important challenge. In recent years, the application of deep learning techniques in solving flow equations has gradually gained widespread attention. Deep learning not only improves solution efficiency by automatically learning complex nonlinear relationships, but also effectively addresses complex problems that are difficult for traditional methods to handle. Existing research indicates that the application of deep learning is primarily focused on the following three directions.

2.1.1. Deep learning-assisted numerical solution methods. Traditional numerical methods often involve large computational costs and limited accuracy, while deep learning can reduce reliance on traditional numerical methods by learning the intrinsic laws of fluid motion, especially in handling nonlinear and complex boundary conditions.

Tim O'Shea[11] points out that physics-informed neural networks (PINN) have become the optimal choice for accelerating the solution of dynamical equations due to their advantages in handling large-scale data and capturing complex nonlinear relationships. Cao et al.[12] proposed a novel framework termed Weight Balanced Physics-Informed Neural Networks (wbPINN), which integrates a multi-component loss formulation with dynamic weighting mechanisms for enhanced multi-task optimization. As illustrated in Figure 1, this methodology incorporates a correlation regularization component and an adaptive penalty term to systematically address interdependencies among primary loss constituents in coupled learning scenarios. Through automated adjustment of loss coefficients via maximum likelihood estimation, the framework achieves balanced gradient propagation across competing objectives, thereby improving both convergence stability and predictive accuracy. Extensive computational evaluations were conducted on two benchmark problems: three-dimensional elliptic equations with intricate geometric interfaces and time-dependent Navier-Stokes systems. Comparative analyses demonstrate that wbPINN outperforms conventional PINN implementations by two orders of magnitude in solution precision while maintaining computational efficiency comparable to state-of-the-art adaptive weighting techniques. Cao et al.[13] proposed a turbulence model correction method based on deep neural networks (DNN) for the supercritical pressure turbulent heat transfer problem. By introducing high-fidelity simulation data, a DNS-DNN-RANS (RANS, Reynolds-averaged Navier-Stokes equations) framework was established to improve the turbulence model for supercritical pressure fluids. This method combines prior physical knowledge and feature engineering strategies, embedding the characteristics of supercritical fluids into turbulence closure, and constructs a regression system suitable for deep learning. The model reproduced heat transfer deterioration and recovery process in vertical pipe upward flow, with a wall temperature prediction error reduced by more than 50% compared to the traditional k-ε model, validating the effectiveness of data-driven methods in turbulence model optimization.

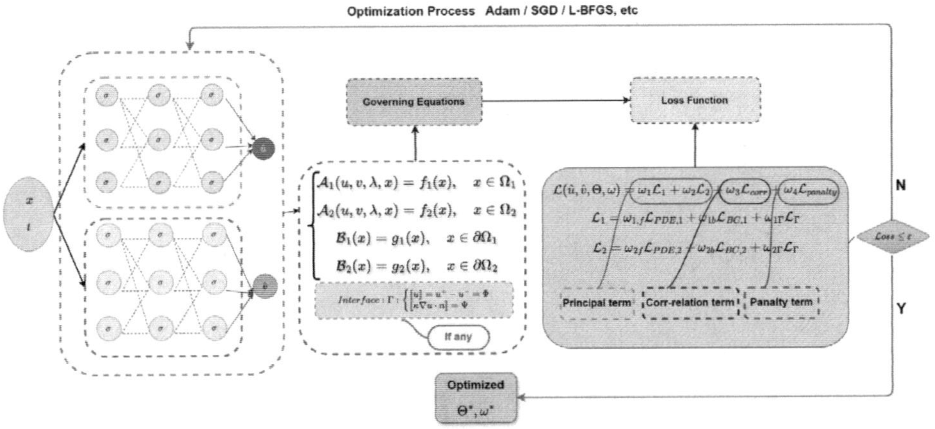

Figure 1. Multi-network architecture based on PINN for solving multi-objective coupled equations[12]

2.1.2. Flow simulation acceleration and real-time prediction. Flow simulations often require extensive computations at multiple time steps and spatial points, especially when dealing with unsteady flows, where the computational load is particularly large. Deep learning models, especially CNN and RNN, can extract spatiotemporal features from historical flow field data through training, enabling rapid flow field prediction. For example, by training a CNN-based model, the evolution of turbulence or shock waves can be predicted in real-time given the initial conditions.

Zhao et al.[14] proposed a deep learning model called unsteady recurrent convolutional neural network (UR-CNN), as shown in Figure 2, for the rapid prediction of unsteady turbulent combustion flow fields in hypersonic vehicle combustion chambers. This model is based on CNN and RNN to extract the spatial distribution characteristics and temporal evolution patterns of the flow field. The results indicate that, under different geometric configurations, the model achieved high computational accuracy, with a correlation coefficient between the predicted results and the actual values exceeding 0.99.

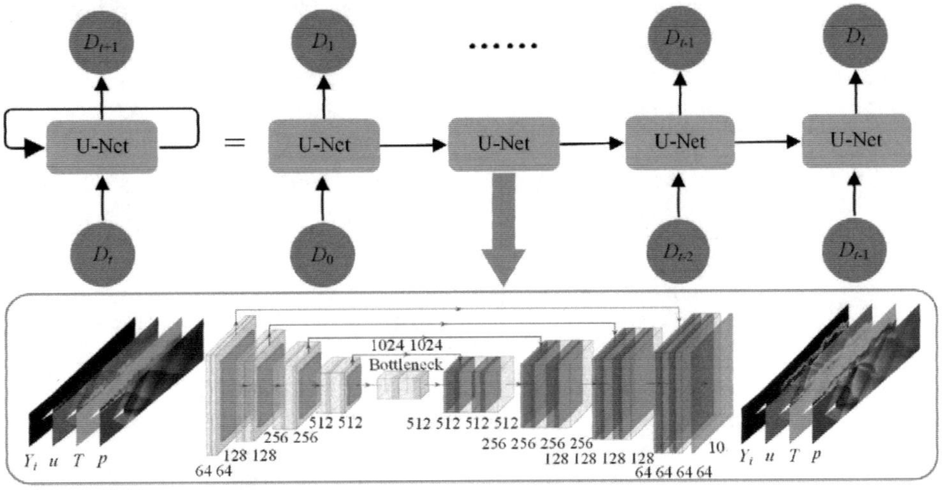

Figure 2. Model structure of UR-CNN[14]

2.1.3. Turbulence and its multiscale modeling. Deep learning applications in turbulence modeling are primarily reflected in its ability to capture and model multiscale phenomena. By using neural networks, turbulence details can be predicted more accurately, thereby improving the precision of turbulence simulations. By using deep neural networks combined with multiscale flow models, both the macroscopic flow characteristics of the fluid and the microscopic vortex effects can be considered simultaneously, thus optimizing the entire computational process. CNN can automatically extract high-frequency features in turbulence through its local receptive fields, thus avoiding the manual parameter setting and complex physical assumptions in traditional turbulence models.

Kai Fukami[15] used deep learning to perform super-resolution analysis on severely under-resolved turbulence field data to reconstruct high-resolution flow fields in his research. As shown in Figure 3, two deep learning models have been developed: CNN and the Hybrid Downsampling Skip Connection / Multiscale (DSC/MS). Other deep learning-based turbulence simulation and prediction studies are shown in Table 1 below.

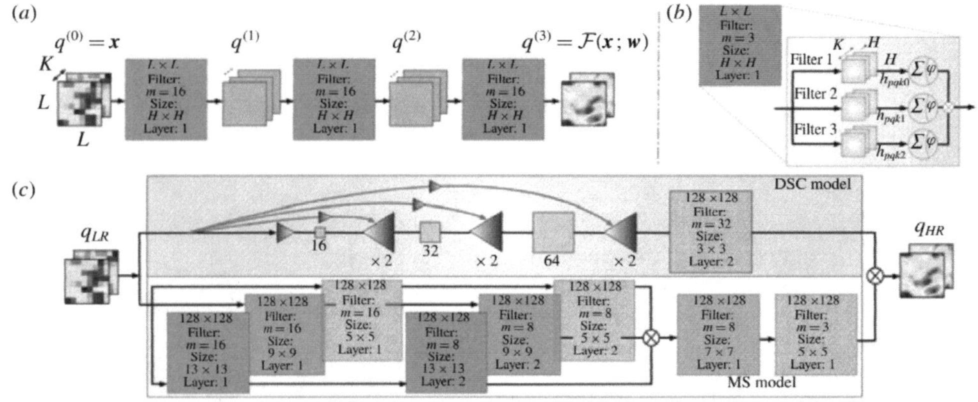

Figure 3. (a) Schematic of CNN with two-dimensional turbulent flow, (b) Internal workings of each CNN, (c) Schematic of DSC/MS [15]

Table 1. Turbulence simulation prediction based on deep learning

Input Parameters	Predicted Parameters	ML Technique(s)	Dataset Size	Result Accuracy	REF
Low-resolution flow field data	High-resolution flow field data	CNN, DSC/MS	Laminar flow: 1000 snapshots Turbulent flow: 10000 snapshots	Achieve high-precision multi-scale reconstruction with low error and accurate statistical properties	[15]
Spatio-temporal data	Physical interpretation	PINNs/DeepONet	Sparse-multi-fidelity	High-precision inversion	[16]
ω, v_t y, ∇	Source terms or their multipliers in the SA model	FNN	CFD simulations: Flat Plate Boundary Layer, Pressure-driven Channel Flow, NACA0012 Airfoil, Onera M6 Wing Transonic Flow	**Interpolation mode:** error < 2% **Extrapolation mode:** accuracy reaches "Good" after introducing diversified data	[17]
u, uu Wuv, Rel	Turbulent kinetic energy production	RANS	Nine DNS/LES datasets with a total of 2,208,503 data points	**Interpolation mode**: $R^2 \approx 87\%$, **Extrapolation mode**: with 5% Hill data,	[18]

term	correlation improves to 80%.

2.2. *Accelerating the solution of mixture physical properties*

The calculation of mixture physical properties, such as density, viscosity, specific heat, and thermal conductivity, plays a crucial role in regenerative cooling system. Traditional computational methods include the use of empirical formulas, equations of state, such as the Redlich-Kwong equation, Peng-Robinson equation. These methods typically require significant computational resources, especially when dealing with multi-component systems, complex phase behavior, and extreme conditions, such as high pressure and high temperature, where their efficiency is relatively low. Deep learning can be applied in the calculation of mixture physical properties, significantly improving both computational efficiency and accuracy.

By training DNN models, the physical properties of a mixture based on parameters such as its components, temperature, and pressure can be obtained quickly. Especially in multi-component systems, neural networks can effectively capture the nonlinear interactions between different components, thereby improving the accuracy of property predictions. Currently, the research on deep learning-based property prediction of supercritical hydrocarbon fuels is summarized in Table 2.

Table 2. Prediction of supercritical hydrocarbon fuel properties based on deep learning

Input Parameters	**Predicted Parameters**	**ML Technique(s)**	**Dataset Size**	**Result Accuracy**	**REF**
T, M, ρ χ_2 χ_3	D_1, D_2, D_3 D_{11}, D_{22}, D_{33}	DNN-TL	1220 sets.Binary mixtures: 80% used for training, 20% used for testing. Ternary mixtures: Only 20% of the data used for training, 80% used for testing	outperforming traditional methods	[14]
P, T	ρ, a, μ, λ Cp, Pr, etc	ANN	49,086 data sets generated by ECS (Pressure 2.5-6 Mpa, Temperature 300-1000 K)	relative errors < 0.1%, and a computational efficiency improvement of 10^4 times	[4]
T, P, ρ Mw, Tc, Pc, ω, etc	D, κ, η, H, S, etc	ANN	Hundreds to thousands of data points	$R^2 > 0.99$ and AARD < 4.7% for both SVR and ANN	[19]

2.3. *Accelerating the solution of chemical reactions*

Chemical reactions of hydrocarbon fuels under supercritical conditions, such as supercritical combustion, supercritical pressure endothermic pyrolysis, oxidation and coking reactions, etc. typically exhibit highly nonlinear, dynamic reaction kinetics and complex multi-physical field interactions, which pose significant challenges to traditional chemical reaction models. Traditional methods, such as computational fluid dynamics (CFD) simulations, although capable of some degree of prediction, are limited in their application due to high computational resource demands and the difficulty in comprehensively describing complex reaction mechanisms.

Deep learning techniques have been used to establish combustion models in recent years, becoming an important method for solving hypersonic combustion chemical reaction calculations. The intelligent improved model showed an average error reduction of over 50% compared to the traditional model. Shahpouri et al.[20] developed two ANN-based models: the 'black-box' model, which is entirely data-driven, and the 'gray-box' model, which combines data-driven approaches with physical mechanisms, for predicting laminar flame speed (LFS). Results indicate that the 'black-box' model is several thousand times faster in terms of computational speed compared to the 'gray-box' model. However, limited research can be found about the application of deep learning techniques in simplification of chenmical reactions in regenerative ccoling processes, which requires further investigations.

2.4. Prediction of macroscopic heat transfer characteristics

Traditional methods, such as CFD and experimental studies, can provide certain accuracy for the heat transfer prediction in regenerative cooling proess, but they have limitations such as high computational cost and limited model scale. Researchers have made progress by combining deep learning algorithms with heat transfer prediction, significantly improving prediction efficiency and accuracy.

Gong et al.[6] proposed a deep neural network model based on multi-layer perceptron (MLP) for predicting the flow field and heat transfer characteristics within the regenerative cooling channels of supercritical hydrocarbon fuels. As shown in Figure 4, by generating training data through CFD, the model is able to predict fuel temperature, velocity, and heat transfer coefficient at a speed 106 times faster than CFD, with a linear correlation coefficient R > 0.99 and a relative error below 2%. The results show that the MLP model successfully captured the effects of inlet Reynolds number, pressure, and heating rate on the flow, providing an efficient tool for the optimization of the regenerative cooling system.

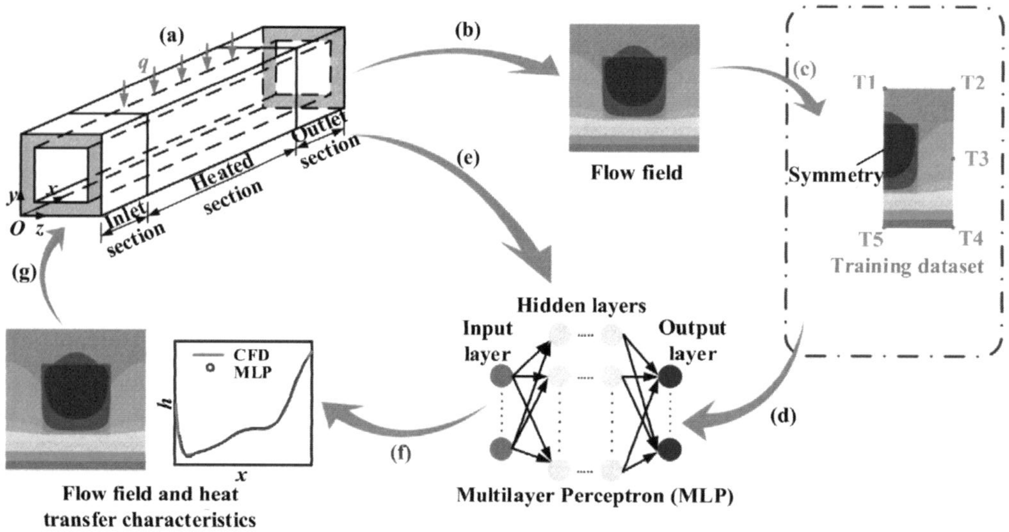

Figure 4. Schematic diagram of flow field and heat transfer prediction based on deep learning[6]

Tao et al.[2] addressed the prediction challenge of Nusselt number (Nu) and friction factor (f) for hydrocarbon fuels under supercritical pressure in horizontal circular tubes by using a Backpropagation (BP) ANN to build a prediction model. Based on experimental data, the network topology was optimized, determining the optimal number of hidden layer neurons to be 9, with the Levenberg-Marquardt algorithm and the tansig transfer function selected. The model predicts the average relative errors of Nu and f to be 4.4% and 3.4%, respectively, over 4917 data points, showing a significant improvement in accuracy compared to traditional empirical correlations, and provides a new approach for supercritical fluid engineering applications.

3. Conclusions and outlooks

Recent progress of deep learning in the simulation of supercritical hydrocarbon fuel flow and heat transfer is given in this paper. Four aspects including flow simulation, mixture property predcition, chemical reaction calculation and heta transfer efficiency prediction were presented and analyzed. The main conclusions are as follows:

(1) For flow simualtion, the application of deep learning is focused on deep learning-assisted numerical solution methods, flow simulation acceleration and real-time prediction and turbulence and multiscale modeling. The physics-informed neural networks can significantly reduces the computational complexity of the Navier-Stokes equations and were deemed to handle large-scale data and capture complex nonlinear relationships.
(2) The physical properties of a mixture can be obtained using the DNN or ANN, with significantly improved computational efficiency and accuracy. DNN can efficiently predict thermophysical properties such as density and viscosity of supercritical fluids. In addtion, DNN exhibits stronger nonlinear fitting capabilities in multicomponent systems.
(3) The deep learning, combining data-driven approaches with physical mechanisms, can rapidly predict the supercritical reaction kinetics parameters. The computational speed is enhanced by thousands of times compared to traditional methods, while maintaining high accuracy. But the current studies only focus on the combustion problem.
(4) The deep learning model based on MLP can predict the heat transfer coefficient at a speed 106 times faster than CFD, with an average relative error of less than 2%, providing an efficient tool for macroscopic heat transfer characteristic prediction in regenerative cooling system.

Research has shown that deep learning has significant advantages in reducing computational costs and improving prediction accuracy. However, its application is still limited by data quality, model generalization capabilities, and the inadequacy of multi-physics coupling mechanisms. Future research is suggested on the improvement of model generalization capabilities, multi-physics coupling modeling, small sample and transfer learning, etc. It is deemed that deep learning will play more and more important role in future engine cooling system design process.

Acknowledgments

The authors appreciate the supports of the National Natural Science Foundation of China (Grant No. 52302478), the National Key Laboratory of Solid Rocket Propulsion (Grant No. SY41YYF202403050).

References

[1] Zhu Y, Peng W, Xu R, et al. Review on active thermal protection and its heat transfer for airbreathing hypersonic vehicles[J]. Chinese Journal of Aeronautics, 2018, 31(10): 1929-1953.
[2] Tao K, Zhu J, Cheng Z, et al. Artificial neural network analysis of the Nusselt number and friction factor of hydrocarbon fuel under supercritical pressure[J]. Propulsion and Power Research, 2022.
[3] Shine S R, Nidhi S S. Review on film cooling of liquid rocket engines[J]. Propulsion and Power Research, 2018, 7(1): 1-18.
[4] Li B, Lee Y, Yao W, et al. Development and application of ANN model for property prediction of supercritical kerosene[J]. Computers & Fluids, 2020, 209: 104665.
[5] Oefelein J C. Thermophysical characteristics of shear-coaxial LOX–H2 flames at supercritical pressure[J]. Proceedings of the Combustion Institute, 2005, 30(2): 2929-2937.
[6] Gong K, Zhang Y, Cao Y, et al. Deep learning approach for predicting the flow field and heat transfer of supercritical hydrocarbon fuels[J]. International Journal of Heat and Mass Transfer, 2024, 219: 124869.
[7] Bazargan M, Mohseni M. Algebraic zero-equation versus complex two-equation turbulence modeling in supercritical fluid flows[J]. Computers & Fluids, 2012, 60: 49-57.
[8] Owoyele O, Kundu P, Ameen M M, et al. Application of deep artificial neural networks to multi-dimensional flamelet libraries and spray flames[J]. International Journal of Engine Research, 2020, 21(1): 151-168.
[9] Selimefendigil F, Öztop H F. Thermal management of nanoliquid forced convective flow over heated blocks in channel by using double elliptic porous objects[J]. Propulsion and Power Research, 2021, 10(3): 262-276.
[10] Şencan A, Kalogirou S A. A new approach using artificial neural networks for determination of

the thermodynamic properties of fluid couples[J]. Energy Conversion and Management, 2005, 46(15-16): 2405-2418.
[11] O'Shea T, Hoydis J. An Introduction to Deep Learning for the Physical Layer[J]. IEEE Transactions on Cognitive Communications and Networking, 2017, 3(4): 563-575.
[12] Cao F, Guo X, Dong X, et al. wbPINN: Weight balanced physics-informed neural networks for multi-objective learning[J]. Applied Soft Computing, 2025, 170.
[13] Cao Y, Xu R, Jiang P. Physics-informed machine learning based RANS turbulence modeling convection heat transfer of supercritical pressure fluid[J]. International Journal of Heat and Mass Transfer, 2023, 201: 123622.
[14] Zhao T, Wu C, Zhu R, et al. A deep learning framework for supersonic turbulent combustion[J]. Acta Astronautica, 2024, 225: 524-537.
[15] Fukami K, Fukagata K, Taira K. Super-resolution reconstruction of turbulent flows with machine learning[J]. Journal of Fluid Mechanics, 2019, 870: 106-120.
[16] Karniadakis G E, Kevrekidis I G, Lu L, et al. Physics-informed machine learning[J]. Nature Reviews Physics, 2021, 3(6): 422-440.
[17] Tracey B D, Duraisamy K, Alonso J J. A machine learning strategy to assist turbulence model development[M]//53rd AIAA Aerospace Sciences Meeting. American Institute of Aeronautics and Astronautics.
[18] Bhushan S, Burgreen G W, Brewer W, et al. Assessment of neural network augmented reynolds averaged navier stokes turbulence model in extrapolation modes[J]. 2023.
[19] Roach L, Rignanese G M, Erriguible A, et al. Applications of machine learning in supercritical fluids research[J]. Journal of Supercritical Fluids, 2023, 202: 106051.
[20] Shahpouri S, Gordon D, Hayduk C, et al. Hybrid emission and combustion modeling of hydrogen fueled engines[J]. International Journal of Hydrogen Energy, 2023, 48(62): 24037-24053.

Journal of Physics: Conference Series **3056** (2025) 012005 doi:10.1088/1742-6596/3056/1/012005

Adaptive Cooperative Attitude Control of Satellite Formation System via Fractional-Order PID

Qixin Kui, Xiaolei Li*, Yinhao Ju, Ruihang Wu, Jiawei Liu.
School of Astronautics, Harbin Institute of Technology, Harbin 150001, China.
E-mail: `xiaoleili@hit.edu.cn`

Abstract. This paper presents a novel robust attitude cooperation scheme that integrates an adaptive cooperative strategy with fractional-order PID for satellite formation systems subject to external disturbances. Firstly, the theoretical foundations for the attitude dynamics model of the considered satellites are derived. Secondly, to achieve attitude cooperation, a cross-coupled attitude cooperative error is designed, where adaptive weighting dynamically regulates cooperative performance among multiple satellites. Thirdly, a fractional-order PID controller is developed based on the adaptive cooperation strategy, which significantly reduces system chattering and improves the attitude cooperation between satellites while ensuring fast attitude convergence. Finally, numerical simulations validate the effectiveness of the presented controllers under disturbance scenarios.

1. Introduction

With the increasing demand for multi-satellite cooperative operations in application domains including celestial exploration and terrestrial monitoring, satellite formation system (SFS) have become a pivotal technology in space applications due to their enhanced spatial resolution and improved system redundancy [1, 2]. However, persistent external disturbances, including solar radiation pressure and gravitational gradient torque in complex space environments, along with attitude instability caused by dynamic interactions between satellites, severely limit the control accuracy and cooperative efficiency of these systems [3, 4]. The critical challenge lies in achieving rapid attitude cooperation among multiple satellites under high-intensity disturbances while ensuring stable formation configuration, which remains a major barrier to the practical deployment of this technology.

Regarding satellite attitude cooperative control, controllers employing sliding mode control theory have demonstrated satisfactory control performance [5–7]. Han et al. developed a novel attitude cooperation strategy combining cross-coupling cooperation with a non-singular fast terminal sliding mode controller for SFS under disturbances, guaranteeing Lyapunov-based convergence stability [7]. Nixon et al. introduced three adaptive continuous sliding mode control algorithms using dual quaternions for SFS, incorporating adaptive gain tuning to address bounded perturbations with unknown magnitudes, and validated their superiority over classical methods via comparative simulations [8]. While traditional satellite attitude control methods achieve basic stabilization, they suffer from critical limitations: PID controllers' fixed gain parameters struggle to adapt to time-varying disturbances, resulting in cooperative error accumulation; although sliding mode control exhibits strong robustness, its inherent high-frequency chattering may intensify inter-satellite coupling oscillations and degrade cooperation accuracy [9].

Recent research efforts have explored the integration of cooperative control theory to enhance cooperative performance in SFS. However, existing strategies often exhibit performance deterioration under abrupt disturbances. Concurrently, advancements in fractional calculus theory have offered novel approaches for spacecraft control, leveraging its inherent memory-dependent characteristics and global modulation capabilities to achieve superior disturbance suppression compared to integer-order controllers [10, 11]. Nevertheless, current studies have yet to fully resolve the integration challenge between fractional-order operators and dynamic cooperative mechanisms, limiting potential performance enhancements.

To overcome these limitations, this paper presents an integrated adaptive cooperative control scheme that combines fractional-order PID (FOPID) control with dynamic cross-coupling cooperation for SFS. First, a cross-coupled attitude cooperative error architecture is designed through ring topology, where adaptive weighting automatically adjusts cooperative performance among multiple satellites under disturbances. This approach enables real-time modulation of cooperation intensity between satellites, effectively reducing cooperative error. Subsequently, a FOPID controller with embedded cooperative adaptation is developed, utilizing fractional calculus operators to improve disturbance rejection while preserving cooperation accuracy. The proposed controller uniquely integrates the memory-dependent properties of FOPID with the adaptive topology's performance to resist nonlinear couplings for controlling satellite attitude cooperation.

The remainder of this paper is organized as follows: The attitude dynamics and kinematics models of the satellite formation system are established in Section 2. In Section 3, the cross-coupled cooperative error design and adaptive weighting are elaborated, followed by the proposal of the FOPID with an adaptive cooperation controller. The effectiveness and superiority of the proposed method are validated through numerical simulations under disturbance scenarios in Section 4. Finally, a concise summary of the research findings is provided in Section 5.

2. System Modeling and Analytical Foundations

2.1. Attitude Dynamics Modeling

This study employs a four-spacecraft formation system (Figure 1) as the research subject. $O_I - X_iY_iZ_i$ represents the Earth-Centered Inertial Reference Frame (F_I). The satellite body-fixed frame, denoted as F_body ($O_B - x_by_bz_b$) , is characterized by attitude quaternionic representation $\boldsymbol{p} = [\ p_0 \quad \boldsymbol{p}_v^{\mathrm{T}}\ ]$ coupled with angular velocity $\boldsymbol{w_b} = [w_{b1}, w_{b2}, w_{b3}]^{\mathrm{T}}$ relative to the inertial frame F_I. The body-fixed coordinate system F_{body} is defined with its origin collocated at the spacecraft's center of mass, with its coordinate axes aligned along the principal axes of the satellite's inertia. Based on this, to describe the attitude motion of the satellite, the following attitude dynamics and kinematics model is established:

$$\begin{cases} \dot{\boldsymbol{p}} = \frac{1}{2}\begin{bmatrix} -\boldsymbol{p_v}^{\mathrm{T}} \\ p_0\boldsymbol{I_{3\times3}} + \boldsymbol{p_v}^{\times} \end{bmatrix}\boldsymbol{w_b} \\ \boldsymbol{J}\frac{\mathrm{d}\boldsymbol{w}_b}{\mathrm{d}t} = -\boldsymbol{w}_b^{\times}\boldsymbol{J}\boldsymbol{w}_b + \boldsymbol{U} + \boldsymbol{D} \end{cases} \tag{1}$$

where $\boldsymbol{J} \in \mathbb{R}^{3\times3}$ denotes the moment of inertia matrix of the satellite, $\boldsymbol{U} \in \mathbb{R}^{3\times3}$ represents the control torque applied to the satellite system, and $\boldsymbol{D} \in \mathbb{R}^{3\times3}$ signifies the external disturbance torque acting on the satellite, which arises from interference factors such as solar pressure and the Earth's gravity gradient torque, $(\bullet)^{\times}$ represents the skew-symmetric matrix of $(\bullet)$.

Additionally, for the convenience of describing the error quaternions and the error angular velocity, the expected reference frame, $F_{ref}(O_R - x_ry_rz_r)$, is defined by an attitude quaternion $\boldsymbol{p}_d = [\ p_{d0} \quad \boldsymbol{p}_{dv}^{\mathrm{T}}\ ]$ and an angular velocity $\boldsymbol{w}_d = [w_{d1}, w_{d2}, w_{d3}]^{\mathrm{T}}$ with respect to F_I. The reference frame F_{ref} shares its origin with F_{body} while maintaining axis alignment with the target orientation. Attitude deviation parameters, comprising the quaternion $\tilde{\boldsymbol{q}}$ and angular rate vector $\tilde{\boldsymbol{w}}$ of F_{body}, are defined relative to F_{ref} as:

$$\tilde{\boldsymbol{p}} = \boldsymbol{p}_d^* \circ \boldsymbol{p} = [p_{d0}, -p_{dv}^{\mathrm{T}}]^{\mathrm{T}} \circ p = [\tilde{p}_0 \quad \tilde{\boldsymbol{p}}_v^{\mathrm{T}}] = \left[p_{d0}p_0 + p_{dv}^T p_v, \left(p_{d0}p_v - p_0 p_{dv} - p_{dv}^{\times} p_v\right)^T\right]^T \tag{2}$$

$$\tilde{\boldsymbol{w}} = w_b - \boldsymbol{E}(\tilde{\boldsymbol{p}})\boldsymbol{w}_d \tag{3}$$

$$\boldsymbol{E}(\tilde{\boldsymbol{p}}) = \left(\tilde{p}_0^2 - \tilde{\boldsymbol{p}}_v^T \tilde{\boldsymbol{p}}_v\right) I_{3\times 3} + 2\tilde{\boldsymbol{p}}_v \tilde{\boldsymbol{p}}_v^T - 2\boldsymbol{p}_0 \tilde{\boldsymbol{p}}_v^{\times} \tag{4}$$

where $\boldsymbol{p}_d^* = [p_{d0} \quad -\boldsymbol{p}_{dv}^{\mathrm{T}}]$ is the conjugate quaternion of $\boldsymbol{p}_d$, and the rotation matrix $\boldsymbol{E}(\tilde{\boldsymbol{p}})$ characterizes the kinematic relationship between F_{ref} and F_{body}. Building on this framework, the fundamental model for attitude error of a satellite is formulated as:

$$\begin{cases} \dot{\tilde{\boldsymbol{p}}} = \dfrac{1}{2} \begin{bmatrix} -\tilde{\boldsymbol{p}}_v^{\mathrm{T}} \\ \tilde{p}_0 \boldsymbol{I}_{3\times 3} + \tilde{\boldsymbol{p}}_v^{\times} \end{bmatrix} \tilde{\boldsymbol{w}} \\ \boldsymbol{J} \dfrac{\mathrm{d}\tilde{\boldsymbol{w}}}{\mathrm{d}t} = \boldsymbol{H} + \boldsymbol{U} + \boldsymbol{D} \end{cases} \tag{5}$$

$$\boldsymbol{H} = -(\tilde{\boldsymbol{w}} + \boldsymbol{E}(\tilde{\boldsymbol{p}})\boldsymbol{w}_d)^{\times} \boldsymbol{J} \left(\tilde{\boldsymbol{w}} + \boldsymbol{E}(\tilde{\boldsymbol{p}})\boldsymbol{w}_d\right) + \boldsymbol{J}\left(\boldsymbol{w}_d{}^{\times} \boldsymbol{E}(\tilde{\boldsymbol{p}})\boldsymbol{w}_d - \boldsymbol{E}(\tilde{\boldsymbol{p}})\dot{\boldsymbol{w}}_d\right) \tag{6}$$

To enhance controller synthesis, the second derivative of the quaternion error vector $\ddot{\tilde{\boldsymbol{p}}}_v$ is analytically obtained through (5), with the associated matrix components formulated as:

$$\boldsymbol{P} = \tilde{p}_0 \boldsymbol{I}_{3\times 3} + \tilde{\boldsymbol{p}}_v^{\times} \tag{7}$$

The angular velocity error vector $\tilde{\boldsymbol{w}}$ is then formulated as $\tilde{\boldsymbol{w}} = 2\boldsymbol{P}^{-1}\dot{\tilde{\boldsymbol{p}}}_v$, and its first derivative is represented as:

$$\dot{\tilde{\boldsymbol{w}}} = 2\dot{\boldsymbol{P}}^{-1}\dot{\boldsymbol{p}}_v + 2\boldsymbol{P}^{-1}\ddot{\boldsymbol{p}}_v \tag{8}$$

Subsequently, neglecting the influence of $\boldsymbol{D}$ on the system, the dynamic equation with the second derivative of $\boldsymbol{p}_v$ as the variable can be established as:

$$\ddot{\tilde{\boldsymbol{p}}}_v = \frac{1}{2}\boldsymbol{P}\left[\boldsymbol{J}^{-1}\boldsymbol{F} - 2\dot{\boldsymbol{P}}^{-1}\dot{\tilde{\boldsymbol{p}}}_v\right] + \frac{1}{2}\boldsymbol{P}\boldsymbol{J}^{-1}\boldsymbol{u} \tag{9}$$

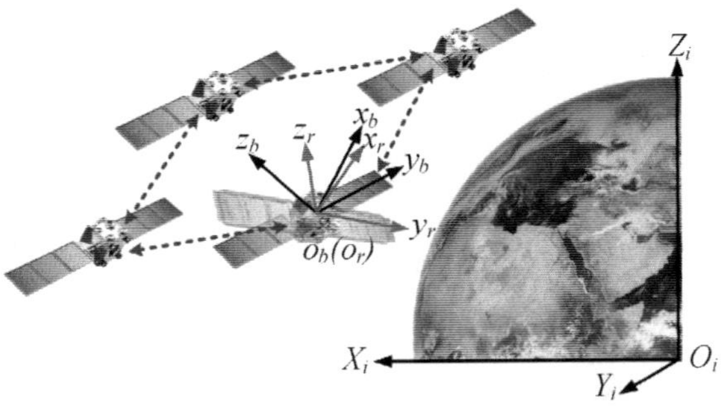

Figure 1. Diagram of the four-satellite formation system

Journal of Physics: Conference Series **3056** (2025) 012005 doi:10.1088/1742-6596/3056/1/012005

2.2. Analytical Foundations

Definition 1 (Riemann-Liouville definition [12]). *For an integrable continuous function $f(t) : \mathbb{R} \to \mathbb{R}$, the α-th order Riemann-Liouville fractional derivative operator is formally defined as:*

$$ {}_{t_0}D_t^{\alpha} f(t) = \frac{1}{\Gamma(n-\alpha)} \frac{d^n}{dt^n} \int_{t_0}^{t} \frac{f(\ell)d\ell}{(t-\ell)^{\alpha+1-n}} \tag{10} $$

where ${}_{t_0}D_t^{\alpha}$ denotes the fractional-order operator, with $n-1 < \alpha < n$ and $0 \leq t_0 < t$, and $\Gamma(\cdot)$ represents the Gamma function defined as:

$$ \Gamma(x) = \int_0^{\infty} e^{-\ell} \ell^{x-1} d\ell \tag{11} $$

3. Attitude Controller

3.1. Adaptive cooperative strategy

An adaptive cooperation scheme has been developed to enhance SFS's cooperative performance. The scheme first derives the tracking error's axis-angle representation through $\tilde{\boldsymbol{p}}_i$, as expressed by:

$$ [\theta_i, x_i, y_i, z_i] = \left[2\arccos \tilde{p}_{i0}, \frac{\tilde{p}_{iv1}}{\sin(\theta_i/2)}, \frac{\tilde{p}_{iv2}}{\sin(\theta_i/2)}, \frac{\tilde{p}_{iv3}}{\sin(\theta_i/2)} \right] \tag{12} $$

where i represents the indexing parameter for constituent spacecraft in SFS.

Subsequently, the attitude error in the form of axial angle is propagated to constituent spacecraft within the i-th node's cross-link network. Utilizing these deviations, the angular discrepancy between the i-th spacecraft and its linked neighbors is computed. Negative values are systematically excluded to mitigate convergence rate degradation induced by cooperative misalignments. This process yields the cooperative error metric θ_{ic} through the following formulation:

$$ \theta_{ic\_left} = \max\{|\theta_i| - |\theta_{i\_left}|, 0\} \quad \theta_{ic\_right} = \max\{|\theta_i| - |\theta_{i\_right}|, 0\} \tag{13} $$

where θ_{i_left} and θ_{i_right} represent the angular deviations of adjacent satellites in communication with satellite i, and θ_{ic} is composed of θ_{ic_left} and θ_{ic_right}. To ensure that the cooperative error can achieve rapid responsiveness across various control phases, a segmented three-stage processing approach is applied to the left and right cooperative errors, respectively. An adaptive function with multi-stage gains for the cooperative error is proposed, which can be expressed as follows:

$$ f_{ic}(\theta_{ic}) = \begin{cases} \lambda_{c1} \cdot \tanh(\theta_{ic}/\theta_{th1}) & \theta_{ic} \geq \theta_{th1} \\ \lambda_{c2} \cdot \left(1 - e^{-k\theta_{ic}}\right) & \theta_{th2} \leq \theta_{ic} < \theta_{th1} \\ \lambda_{c3} & \theta_{ic} < \theta_{th2} \end{cases} \tag{14} $$

where θ_{th1} and θ_{th2} represent different stages of error angle, $\lambda_{c1}, \lambda_{c2}$ and λ_{c3} signify constant gains at different stages, k denotes the constant coefficient. Then, weights are assigned based on the absolute values of the cooperative errors from the left and right neighboring satellites. The dynamic weights can be described as follows:

$$ m_{i\_c} = \frac{|\theta_i - \theta_{i\_c}|}{\sum\limits_{k \in \mathcal{N}_i} |\theta_i - \theta_k| + \varepsilon} \tag{15} $$

where $\mathcal{N}_i$ represents the set of adjacent satellites of satellite i, ε represents a minimal constant introduced to avoid division by zero. The adjusted cooperative tracking error, incorporating dynamic weight allocation, can be denoted as follows:

$$ \theta_{i*} = m_{i\_left} \cdot f_{ic}(\theta_{ic\_left}) + m_{i\_right} \cdot f_{ic}(\theta_{ic\_right}) \tag{16} $$

Journal of Physics: Conference Series **3056** (2025) 012005 doi:10.1088/1742-6596/3056/1/012005

The adjusted cooperative tracking error can be expressed as:

$$\theta_i^* = \theta_i + \theta_{i^*} \tag{17}$$

To ensure convergence speed and accuracy while avoiding singularities, the tracking error is ultimately subjected to amplitude limiting. The final adjusted cooperative tracking error can be described as follows:

$$\theta_i^{**} = \max\left\{\min\left\{\theta_i^*, \pi/2\right\}, 0\right\} \tag{18}$$

Based on the tracking error axis $[x_i, y_i, z_i]$ and the final adjusted cooperative tracking error θ_i^{**}, the cooperative attitude error can be mathematically represented by the quaternion:

$$\tilde{\boldsymbol{p}}_{ci} = [\tilde{p}_{ci0} \quad \tilde{\boldsymbol{p}}_{civ}^{\mathrm{T}}] = [\cos(\theta_i^{**}/2, x_i), \sin(\theta_i^{**}/2), y_i \sin(\theta_i^{**}/2), z_i \sin(\theta_i^{**}/2)] \tag{19}$$

3.2. Fractional-order PID attitude cooperative controller

To facilitate the control of the satellite, the integrated error is defined as:

$$\boldsymbol{E}_i = \tilde{\boldsymbol{p}}_{iv} + \boldsymbol{w}_i \tag{20}$$

Based on the integrated error and **Definition 1**, the fractional-order PID controller can be represented as: The fractional-order PID controller can be represented as

$$\boldsymbol{u}_{ai} = c_{pai}\boldsymbol{E}_i + c_{iai}D^{-\alpha}\boldsymbol{E}_i + c_{dai}D^{\alpha}\boldsymbol{E}_i \tag{21}$$

where c_{pai}, c_{iai} and c_{dai} denote constant coefficients. Combining the cooperative error with the integrated error, the integrated cooperative error can be expressed as:

$$\boldsymbol{E}_i^* = \tilde{\boldsymbol{p}}_{ifv} + \tilde{\boldsymbol{w}}_i \tag{22}$$

where $\tilde{\boldsymbol{q}}_{ifv}$ can be represented by the following adaptive function:

$$\tilde{\boldsymbol{p}}_{ifv} = \begin{cases} \tilde{\boldsymbol{p}}_{iv} = [x_i \sin(\theta_i/2),\ y_i \sin(\theta_i/2),\ z_i \sin(\theta_i/2)]^{\mathrm{T}} & \theta_i \le \theta_{i\_\min} \\ \tilde{\boldsymbol{p}}_{civ} = [x_i \sin(\theta_i^{**}/2),\ y_i \sin(\theta_i^{**}/2),\ z_i \sin(\theta_i^{**}/2)]^{\mathrm{T}} & \theta_i > \theta_{i\_\min} \end{cases} \tag{23}$$

Based on the integrated cooperative error and **Definition 1**, a fractional-order PID attitude cooperative controller can be designed as

$$\boldsymbol{u}_{bi} = c_{pbi}\boldsymbol{E}_i^* + c_{ibi}D^{-\alpha}\boldsymbol{E}_i^* + c_{dbi}D^{\alpha}\boldsymbol{E}_i^* \tag{24}$$

where c_{pbi}, c_{ibi} and c_{dbi} denote constant coefficients.

4. Simulation Experiments

To validate the efficacy and improvement of the designed fractional-order PID attitude cooperative controller and adaptive cooperative strategy, simulation experiments are established. Following the dynamic and kinematic frameworks outlined in Section 2, a four-satellite formation system is constructed. Each satellite's rotational inertia matrix is configured as:

$$\boldsymbol{J}_i = \begin{bmatrix} 20 & 1.2 & 0.9 \\ 1.2 & 17 & 1.4 \\ 0.9 & 1.4 & 15 \end{bmatrix} \ \mathrm{kg}\cdot\mathrm{m}^2$$

The initial angular velocities for all constituent satellites within the SFS are specified as follows:

$$w_i(0) = \begin{bmatrix}0.06 & -0.04 & 0.05\end{bmatrix}^{\mathrm{T}} \mathrm{rad}\cdot\mathrm{s}^{-1}$$

The initial quaternion states for all formation members are prescribed as:

$$p_1(0) = [0.9764 \quad 0.1209 \quad 0.1776 \quad 0.0199]^{\mathrm{T}} \quad p_2(0) = [0.9945 \quad 0.0396 \quad 0.0889 \quad 0.0396]^{\mathrm{T}}$$
$$p_3(0) = [0.9725 \quad 0.2117 \quad 0.0944 \quad 0.0236]^{\mathrm{T}} \quad p_4(0) = [0.9811 \quad -0.0151 \quad 0.1730 \quad 0.0858]^{\mathrm{T}}$$

The initial values of the quaternion for the target attitude of each satellite are selected as:

$$p_{1,t}(0) = [0.9020 \quad 0.2376 \quad 0.3603 \quad 0.0150]^{\mathrm{T}} \quad p_{2,t}(0) = [0.9710 \quad 0.1050 \quad 0.1875 \quad 0.1050]^{\mathrm{T}}$$
$$p_{3,t}(0) = [0.9380 \quad 0.0708 \quad 0.2028 \quad 0.2720]^{\mathrm{T}} \quad p_{4,t}(0) = [0.9443 \quad 0.2335 \quad 0.1380 \quad 0.1862]^{\mathrm{T}}$$

which is governed by the angular rate vector through the following relation:

$$w_{i,d}(0) = \left[\frac{1}{20}\sin(\frac{\pi t}{100}),\ \frac{1}{20}\sin(\frac{\pi t}{50}),\ \frac{1}{20}\sin(\frac{3\pi t}{100})\right]\ \mathrm{rad}\cdot\mathrm{s}^{-1}$$

To precisely account for inter-satellite link constraints and external disturbances, a 140-millisecond propagation latency is incorporated across adjacent spacecraft nodes, with the disturbance for individual satellites configured parametrized as:

$$\boldsymbol{D}_i = \left[\frac{3}{100}\sin(t),\ \frac{3\pi t}{50}\sin(\frac{6t}{5}),\ \frac{9\pi t}{100}\sin(\frac{3t}{2})\right]^{\mathrm{T}}\ \mathrm{N}\cdot\mathrm{m}$$

Furthermore, a cooperative performance function is introduced to assess the attitude control effectiveness within the SFS, mathematically described as::

$$V_c = \sqrt{\sum\nolimits_{i=1}^{4}\sum\nolimits_{k=1}^{3}\tilde{p}_{ik}^2}$$

where k denotes the index of elements in $\tilde{p}_i$. A lower value of V_c indicates superior cooperation performance within the SFS.

The numerical experiment is configured with a 10-second duration and 20-millisecond time step. A comparative analysis is conducted employing four distinct control strategies: conventional PID, nonlinear fast terminal sliding mode (NFTSM) [7], fractional-order PID (FOPID), and FOPID with adaptive cooperation (FOPID with AC). The corresponding control parameters for these methodologies are detailed in Table 1.

Table 1. Parameters of four controllers

Controllers	**Control Parameters**
PID	$K_p = 40,\ K_i = 20,\ K_d = 1.$
NFTSM	$\alpha = \beta = 0.5,\ g = 7,\ h = 3,\ p = 5,\ q = 3,\ m = 1,\ n = 3,$ $\theta_{\max} = \pi,\ \theta_{\min} = 0.05, \rho_1 = \rho_2 = 1.$
FOPID	$c_{pai} = 120,\ c_{iai} = 5,\ c_{dai} = 1.$
FOPID with AC	$c_{pbi} = 120,\ c_{ibi} = 5,\ c_{dbi} = 1,\ \lambda_{c1} = 5,\ \lambda_{c2} = 2,\ \lambda_{c3} = 0.02,$ $\theta_{ih1} = 0.5\pi,\ \theta_{ih2} = 0.05\pi,\ \varepsilon = 10^{-6}, \alpha = 0.5,\ \theta_{i\_\min} = 0.05.$

The simulation results are shown in Figure 2 to Figure 5. Figure 2 presents the Euler angle error trajectories of Satellite 3, demonstrating the control performance of the FOPID with AC strategy across all degrees of freedom of the satellite's attitude. The three curves represent the pitch, roll, and yaw errors of the satellite. Under the FOPID with AC strategy, the FOPID with AC controller smoothly adjusts the attitude angles to the target orientation. As shown in Figure 2, all angular errors converge to near zero within 3 seconds, which confirms the effectiveness of the FOPID with AC.

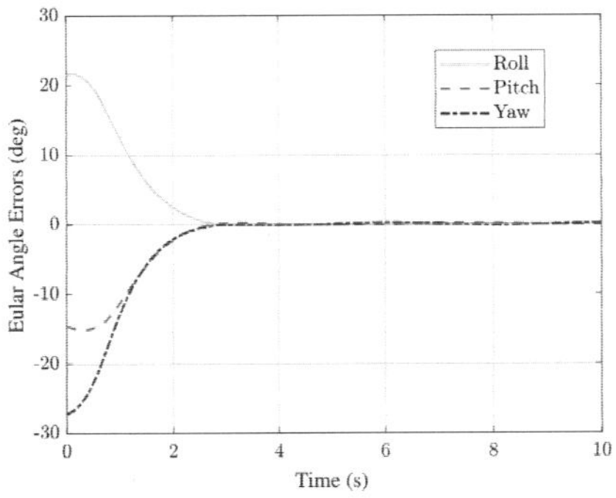

Figure 2. Trajectories of θ_3 under FOPID with AC.

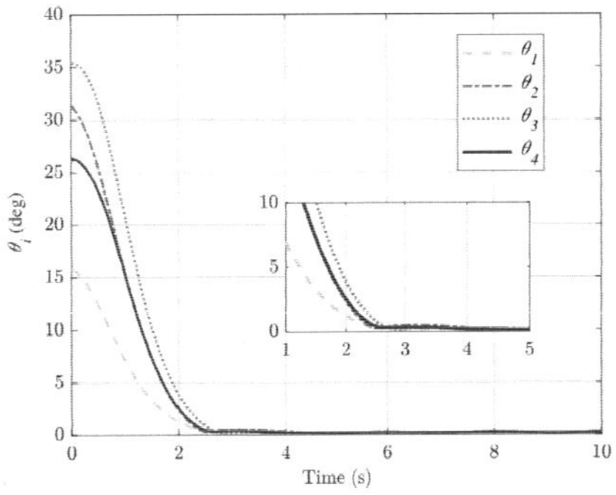

Figure 3. Trajectories of θ_i under FOPID with AC.

Figure 3 depicts the error angles of attitude control for four satellites, demonstrating the overall control effectiveness of the FOPID with AC control strategy on each satellite's attitude. Evidently, the proposed FOPID with AC exhibits significant advantages in both regulation speed and cooperation. As shown in Figure 3, the FOPID with AC controller smoothly regulates all error angles toward zero, and all error angles converge simultaneously to near-zero values within 3 seconds, confirming the rapidity and cooperation performance of the FOPID with AC.

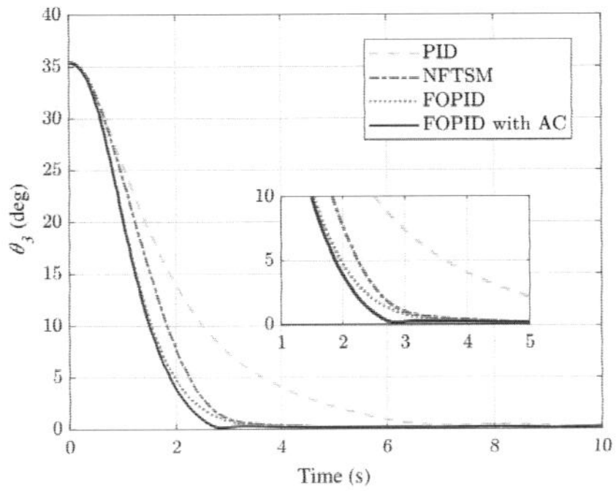

Figure 4. Trajectories of θ_3 under different controllers.

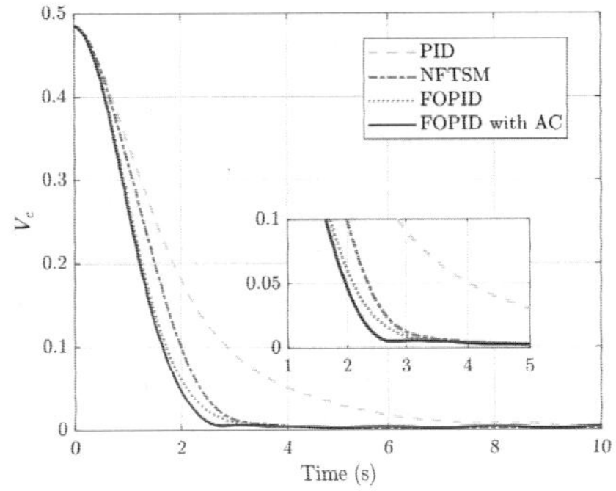

Figure 5. Trajectories of V_c under different controllers.

The comparative control characteristics of the SFS under multiple methodologies are shown in Figures 4 and 5, demonstrating the superiority of the FOPID with AC. Figure 4 shows the attitude tracking error convergence trajectories of Satellite 3 under various control methods. Compared with PID and NTSMC methods, the proposed control method achieves significant improvement in rapidity, with further enhancements over the FOPID. The FOPID with AC exhibits the fastest convergence speed. To comparatively evaluate controller cooperation,

Figure 5 depicts the cooperative performance (V_c) profiles under four distinct control strategies. Evidently, the V_c of the SFS under FOPID with AC converges most rapidly to near-zero values, demonstrating the cooperation superiority of the proposed method. These experimental results systematically confirm both the accelerated response and superior cooperation capabilities of the proposed method.

5. Conclusions

In this paper, a novel adaptive robust cooperative control scheme is presented, which achieves adaptive attitude cooperation of SFS under disturbances. The scheme comprises two contributions: (1) An adaptive cooperative strategy is proposed, which constructs a novel attitude coordination error based on cross-coupled topology, with an adaptive weighting mechanism that effectively regulates cooperative performance among multiple satellites. (2) A fractional-order PID controller with an adaptive cooperative strategy is developed, which is built upon fractional calculus theory and significantly reduces system chattering while enhancing the cooperative performance in SFS. Numerical simulations show the proposed strategy outperforms conventional methods in convergence speed, attitude cooperation accuracy, and disturbance rejection, demonstrating its effectiveness and superiority.

Acknowledgments

This work was supported in part by the National Natural Science Foundation of China under Grant 62303139, in part by the Science Center Program of National Natural Science Foundation of China under Grant 62188101, and in part by the Heilongjiang Touyan Team Program.

References

[1] D. Selva, A. Golkar, O. Korobova, I. L. I. Cruz, P. Collopy and O. L. de Weck, "Distributed earth satellite systems: What is needed to move forward", in Aerospace Information Systems, vol. 14, no. 8, pp. 412-438, 2017.

[2] L. Cao, Y. Chen, Z. Zhang, H. Li and A. K. Misra, "Predictive Smooth Variable Structure Filter for Attitude Synchronization Estimation During Satellite Formation Flying," in IEEE Transactions on Aerospace and Electronic Systems, vol. 53, no. 3, pp. 1375-1383, June 2017.

[3] R. Kristiansen and P. J. Nicklasson, "Spacecraft formation flying: A review and new results on state feedback control. ACTA", Astronautica, vol. 65, no. 11, pp. 1537-1552, 2009.

[4] X. Li, Y. Ju and X. Wang, "Practical Attitude Cooperative Control of Satellite Formation System Via Fuzzy-Based Fixed-Time Sliding Mode," IEEE Transactions on Aerospace and Electronic Systems, doi: 10.1109/TAES.2025.3545397.

[5] X. Li, G. Sun, S. Han and X. Shao, "Fractional-Order Nonsingular Terminal Sliding Mode Tension Control for the Deployment of Space Tethered Satellite," IEEE Transactions on Aerospace and Electronic Systems, vol. 57, no. 5, pp. 2759-2770, 2021.

[6] X. Tao, F. Zhang, B. Huang, G. Shen and P. Huang, "Fixed-Time Sliding Mode Coordinated Deployment Control for Space Triangle Tethered Formation System," in IEEE Transactions on Aerospace and Electronic Systems, vol. 60, no. 2, pp. 1351-1363, April 2024.

[7] S. Han, Y. Ju, S. Chen, F. Tang, Y. Wang and G. Sun, "Cross-coupling Attitude Synchronous Control of Satellite Formation via Fast Terminal Sliding Mode," 2024 36th Chinese Control and Decision Conference (CCDC), Xi'an, China, 2024, pp. 4736-4741.

[8] M. E. Nixon and Y. B. Shtessel, "Adaptive Sliding Mode Control of a Perturbed Satellite in a Formation Antenna Array," in IEEE Transactions on Aerospace and Electronic Systems, vol. 58, no. 5, pp. 4595-4614, Oct. 2022.

[9] Y. Shtessel, C. Edwards, L. Fridman and A. Levant, Sliding Mode Control and Observation, New York, NY, USA, Springer-Verlag, 2014.

[10] X. Li, G. Sun, Z. Kuang, S. Han, "Nonlinear Predictive Optimization for Deploying Space Tethered Satellite via Dis-crete-time Fractional-Order Sliding Mode," IEEE Transactions on Aerospace and Electronic Systems, 58 (5): 4517-4526, 2022.

[11] X. Li, G. Sun, C. Xue, "Fractional-Order Deployment Control of Space Tethered Satellite via Adaptive Su-per-Twisting Sliding Mode," Aerospace Science and Technology, 121: 107390, 2022.

[12] I. Podlubny, Fractional Differential Equations, Academic, San Diego, CA, USA, 1999, pp. 243259.

Graph-Based Trajectory Planning for Spacecraft within Multiple High-Speed Debris Conjunctions

Jikai Wang, Xiaoyi Wang* and Ming Xu

School of Astronautics, Beihang University, Beijing 100191, China

E-mail: `xiaoyi.wang@buaa.edu.cn (corresponding author)`

Abstract. Because of the great increase in space debris, collision risks for active satellites are heightened. Existing avoidance methods, which focus on isolated conjunctions, are insufficient, especially when there are multiple high-speed debris conjunctions. In this study, we propose a graph-based trajectory planning framework for the systematic search of safe paths in complex space debris environments. Clohessy-Wiltshire equations are used to model the maneuvering trajectory of spacecraft. Velocity and time are incorporated as grid search variables. The debris obstacle space is defined to guarantee the safety and feasibility of the trajectory. An analytical expansion strategy is proposed to improve efficiency and avoid time-wasting exploration. Three priority-based search strategies are developed: fuel-optimal, time-depth priority, and heuristic-guided strategies. Simulations within different scenarios confirm the method's ability to generate safe, feasible trajectories. In the results, the heuristic-guided strategy achieves near-optimal fuel consumption with greatly reduced computational time cost. The proposed approach provides effective and scalable solutions for autonomous spacecraft collision avoidance and future space traffic management.

1. Introduction

The development of space technology and the deployment of mega-constellations such as Starlink have greatly increased the number of space objects[1]. Low Earth orbit is becoming increasingly crowded, with lots of satellites and space debris densely distributed around the planet. This explosive growth significantly increases the likelihood of the Kessler syndrome[2], which makes frequent collision avoidance maneuvers (CAMs) essential for satellites' daily operations.

Traditionally, CAM planning focused on impulsive maneuvers aimed at maximizing miss distance or minimizing collision probability with minimal fuel cost. Bombardelli and Hernado-Ayuso[3] proposed an analytical method based on the linear relationship between impulsive maneuver and the resultant displacement on the b-plane, effectively reducing the optimization problem to an eigenvalue formulation, thus improving computational efficiency for CAM planning. Similarly, Gonzalo et al.[4] leverages Gauss's planetary equation and linearized relative motion models to analytically map maneuver effects to deviations at close approach. Morselli et al.[5] formulated CAM design as a multi-objective optimization problem, employing multi-objective particle swarm optimization (MOPSO) to simultaneously minimize fuel consumption, maximize miss distance, and manage mission constraints across multiple potential conjunctions. On the other hand, low-thrust CAM strategies are receiving growing attention due to the widespread adoption of electric propulsion systems. Ströbel and Stoll[6] developed an analytical method to design low-thrust CAMs efficiently, exploiting Hill-Clohessy-Wiltshire(CW) dynamics

Journal of Physics: Conference Series **3056** (2025) 012006 doi:10.1088/1742-6596/3056/1/012006

to enable fast onboard implementation and robust handling of long-duration thrusting profiles. Hernado-Ayuso and Bombardelli[7] formulated the optimal control problem for continuous low-thrust collision avoidance based on B-plane dynamics and solved it using an indirect method, achieving high precision in the maneuver design.

With increasing debris density, simply avoiding individual objects can create secondary collision risks. This highlights the pressing need for advanced multi-conjunction strategies to maintain operational safety and sustainability. Kim et al.[8] proposed a genetic algorithm (GA) framework to optimize collision avoidance maneuvers under multiple debris threats, considering constraints such as fuel consumption and allowable collision probability. In their work, Nag et al.[9] formulated the problem of resolving multiple potential collisions as a Markov Decision Process (MDP) and applied a greedy algorithm to compute feasible maneuver solutions. Masson et al.[10] formulated the problem as a non-convex quadratic optimization, introducing operational constraints such as station-keeping and predefined maneuver execution windows. They demonstrated that by employing advanced numerical methods, including semi-definite programming and branch-and-bound algorithms, it is possible to balance maneuver optimality and computational feasibility for complex multi-threat scenarios. Similarly, Pavanello et al.[11, 12] developed an innovative sequential convex programming approach combined with second-order cone programming and differential algebra techniques.

Existing methods often rely on iterative optimization from initial guesses, which risks entrapment in local optima or failure to find feasible solutions due to limited spacecraft maneuverability. In contrast, graph search algorithms, widely used in autonomous driving[13] and UAV path planning[14], offer a systematic and efficient global search approach. Safe spacecraft navigation in complex debris environments is analogous to vehicle navigation through cluttered spaces, both seeking collision-free, energy-efficient paths.

However, unique challenges in spacecraft dynamics and collision assessment distinguish this problem. To address these, this study proposes a graph search algorithm that incorporates spacecraft dynamic constraints for reliable planning in dense debris fields. The main contributions are shown as follows. (1) The state space is discretized with each node representing a continuous state range, and time is explicitly introduced as a variable. (2) A collision detection criterion based on instantaneous conjunction assumptions evaluates debris risks. (3) To improve efficiency, an analytic expansion strategy accelerates convergence by pruning unnecessary exploration. Finally, three search strategies are designed and compared across various debris scenarios to evaluate performance.

2. Scenario Description and Dynamic Model

As the number of orbital objects in space continues to grow, the frequency of close conjunction events has increased significantly, requiring spacecraft to perform frequent collision avoidance maneuvers to ensure safe operations. However, with the near-Earth space becoming increasingly congested, maneuvering to avoid one piece of debris may inadvertently introduce new collision risks with other objects, as illustrated in Fig.1. Traditional single-debris avoidance algorithms are insufficient for planning an optimal trajectory in such environments. A more advanced algorithm is required, one that can simultaneously account for multiple debris threats to generate a globally optimal avoidance path.

Fig.2 illustrates the relationship between the spacecraft's maneuver trajectory and the flyby paths of space debris, as represented in the virtual reference orbital coordinate frame. The origin of this frame is located at the center of mass of the virtual reference spacecraft. The X-axis is aligned with the radial direction pointing from the Earth's center, the Y-axis is aligned with the direction of orbital velocity, while the Z-axis completes the right-handed coordinate system and points along the orbital angular momentum vector, normal to the orbital plane. After executing a maneuver, the spacecraft's relative motion with respect to the reference orbit

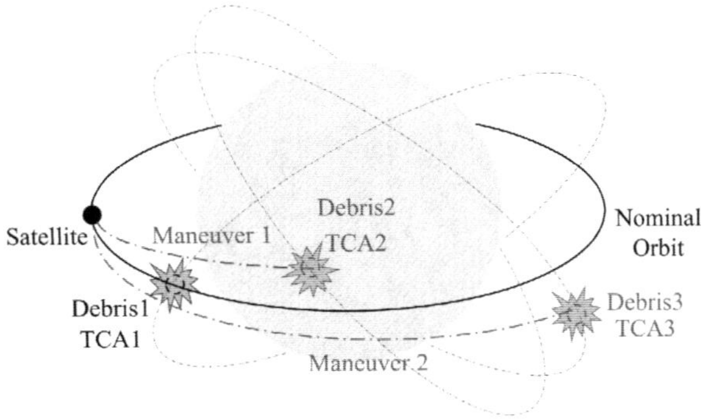

Figure 1: Multi-Conjunction Collision Scenario.

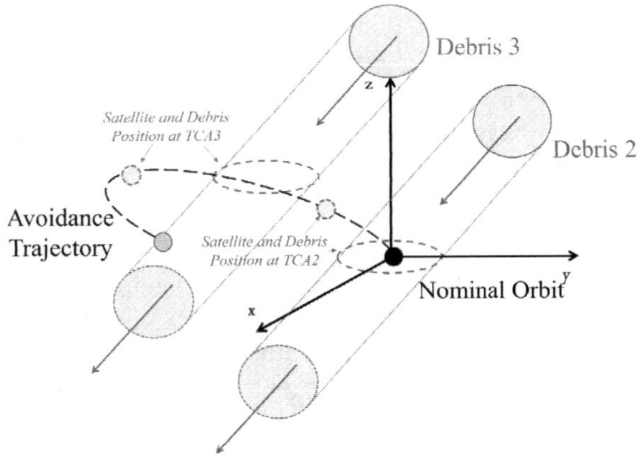

Figure 2: Spacecraft Collision Avoidance Maneuver.

can be described by the state vector $\boldsymbol{x} = [\boldsymbol{r}^T, \boldsymbol{v}^T]$, where $\boldsymbol{r}$ represent the relative position, and $\boldsymbol{v}$ represent the corresponding relative velocity. The spacecraft's motion follows the Clohessy-Wiltshire equations, and the dynamic model can be expressed as:

$$\begin{cases} \dot{\boldsymbol{x}} = \boldsymbol{A}\boldsymbol{x} + \boldsymbol{B}\boldsymbol{u} \\ \boldsymbol{A} = \begin{bmatrix} 0 & 0 & 0 & 1 & 0 & 0 \\ 0 & 0 & 0 & 0 & 1 & 0 \\ 0 & 0 & 0 & 0 & 0 & 1 \\ 3n^2 & 0 & 0 & 0 & 2n & 0 \\ 0 & 0 & 0 & -2n & 0 & 0 \\ 0 & 0 & -n^2 & 0 & 0 & 0 \end{bmatrix}, \boldsymbol{B} = \begin{bmatrix} 0 & 0 & 0 \\ 0 & 0 & 0 \\ 0 & 0 & 0 \\ 0 & 0 & 0 \\ 1 & 0 & 0 \\ 0 & 1 & 0 \\ 0 & 0 & 1 \end{bmatrix} \end{cases} \tag{1}$$

where $n = \sqrt{\mu/a^3}$ is the orbital angular velocity of the spacecraft, $\mu = 3.986 \times 10^{14} m^3/s^2$ is the gravitational constant of Earth, and a denotes the semi-major axis of the reference orbit.

The spacecraft's state after an impulsive collision avoidance maneuver can be analytically computed using the state transition matrix, as follows:

$$\boldsymbol{x}(t+\tau) = \boldsymbol{\phi}(\tau)(\boldsymbol{x}(t+\tau) + \delta\boldsymbol{v}) \tag{2}$$

where $\delta\boldsymbol{v} = [0, 0, 0, \delta v_x, \delta v_y, \delta v_z]^T$ represents the instantaneous velocity increment due to the impulsive maneuver. The state transition matrix $\boldsymbol{\phi}(\tau)$ is derived analytically from the CW equations and has a fixed form, which is given by:

$$\phi(\tau) = \begin{bmatrix} 4-3\cos(n\tau) & 0 & 0 & \frac{1}{n}\sin(n\tau) & \frac{2}{n}(1-\cos(n\tau)) & 0 \\ 6(\sin(n\tau)-n\tau) & 1 & 0 & \frac{2}{n}(\cos(n\tau)-1) & \frac{4}{n}\sin(n\tau) & -3\tau \\ 0 & 0 & \cos(n\tau) & 0 & 0 & \frac{1}{n}\sin(n\tau) \\ \frac{3n}{2}\sin(n\tau) & 0 & 0 & \cos(n\tau) & 2\sin(n\tau) & 0 \\ 6n(\cos(n\tau)-1) & 0 & 0 & -2\sin(n\tau) & 4\cos(n\tau)-3 & 0 \\ 0 & 0 & -n\sin(n\tau) & 0 & 0 & \cos(n\tau) \end{bmatrix} \quad (3)$$

The debris considered in this study is characterized by an extremely high relative velocity at the time of closest approach (TCA). To account for the positional uncertainties of the debris, a simplified model is adopted in which a spherical safety zone with radius $D_{\text{threshold}}$ is defined around the nominal debris position. At the moment of conjunction, this spherical safety region sweeps out a cylindrical volume in the O-XYZ coordinate frame. Its projection onto the O-XY plane forms an elliptical area, as depicted in Fig. 2, which represents the obstacle space for the spacecraft. It should be noted that this debris model is a simplified approach. But, it serves the purpose of validating the effectiveness of the planning method. In practice, the $D_{\text{threshold}}$ is adjustable according to the prediction uncertainties. We treat debris as non-maneuvering objects on predictable paths. The safety radius is assumed large enough to account for potential positional uncertainties.

Given the very short duration of the conjunction event, the spacecraft's motion during the TCA can be reasonably neglected. Therefore, the collision risk is considered negligible if the spacecraft remains outside the obstacle space at the corresponding TCA. The objective of the trajectory planning problem for avoiding multiple high-speed debris encounters is to determine a discrete control sequence $[\boldsymbol{u}_0, \boldsymbol{u}_1, ..., \boldsymbol{u}_n]$ that generates a trajectory ensuring the spacecraft remains outside the obstacle space of all debris objects at their respective TCAs.

Compared with conventional path planning problems, the multi-conjunction collision avoidance problem for spacecraft involves significantly more complex dynamics and cannot be simplified as a two-dimensional grid search. Both position and velocity must be incorporated as state variables in the search process. Moreover, due to the high relative velocities of space debris, which act as highly dynamic obstacles, time must also be included in the planning framework. The spacecraft's limited maneuvering capability imposes stringent constraints on its control authority, rendering fuel consumption, rather than path length, the primary optimization objective. In addition, conventional collision metrics based on Mahalanobis or Euclidean distances are no longer applicable and must be replaced with scenario-specific conjunction evaluation criteria. Given these challenges, a novel graph-based search framework is required to address the unique requirements of spacecraft collision avoidance trajectory planning.

3. Algorithm Design

To achieve spacecraft collision avoidance within multiple conjunction scenarios, a novel algorithmic framework is developed, which integrates orbital dynamics modeling, debris collision risk assessment, and maneuver feasibility evaluation within a graph search algorithm. By embedding these elements into the search process, the proposed method ensures collision safety, physical feasibility, and computational efficiency. The framework is well-suited for spacecraft collision avoidance tasks in complex scenarios involving multiple debris objects with potential collisions. The following sections provide a detailed description of the algorithm's implementation.

3.1. Grid Map Design

For typical graph search problems, the first step involves discretizing the dynamic space to construct a grid map. To ensure trajectory feasibility and better accommodate complex

Journal of Physics: Conference Series **3056** (2025) 012006 doi:10.1088/1742-6596/3056/1/012006

environments, this study adopts the continuously variable grid node approach proposed by Dolgov[13]. In this method, each grid cell is associated with a continuous range of the spacecraft's motion state at a given time, as illustrated in Fig.3. Unlike traditional grid discretization methods, where each grid corresponds to a fixed state, in this approach, each grid cell represents a continuous subset of the state space. If the spacecraft's state falls within the corresponding range, it is considered to have reached that grid node. This design ensures that the resulting search paths inherently satisfy the underlying dynamic constraints.

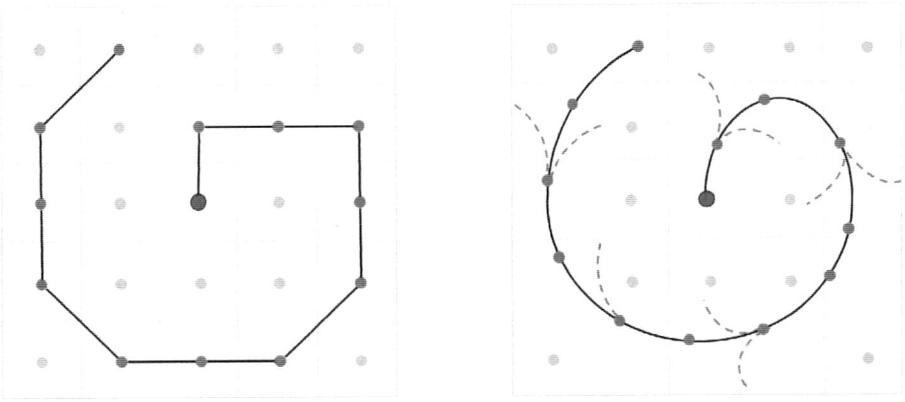

Figure 3: Schematic Diagram of Grid Discretization.

In contrast to traditional static obstacle avoidance problems, space debris exhibits highly dynamic behavior relative to the spacecraft, with conjunctions occurring at specific times. Consequently, in addition to position and velocity, time must also be incorporated as a variable in the search space. For this problem, the search space is defined as $(x, y, z, v_x, v_y, v_z, t)$, where x, y, z represent the relative position, v_x, v_y, v_z the relative velocity of the spacecraft with respect to the nominal orbit in the RTN coordinate frame, and t denotes the time. While position and velocity are treated as continuous variables, time is discretized into multiple fixed steps according to a chosen time increment. A coarse grid cannot distinguish between different paths, thus leading to planning failure. A fine grid leads to excessive node expansion, reducing efficiency. The optimal resolution requires case-specific tuning. But this is beyond the scope of this paper.

3.2. Algorithm Design

The framework of the graph search algorithm proposed in this study is outlined in Algorithm 1. Starting from the initial node, the algorithm iteratively explores all possible maneuvering decisions to generate new nodes. Each node is assigned a priority value, and a finite-priority queue is maintained to manage the order of node expansions. At each iteration, the node with the highest priority is extracted from the queue, and its neighboring nodes are generated based on the current state of this node. If a newly generated node already exists in the search space, its state is updated only if the new path offers a higher priority value.

Furthermore, any nodes identified by the `CollisionCheck` function as either colliding with debris or violating the predefined configuration boundaries are pruned and excluded from further exploration. Additionally, through the `AnalyticalExpansion` function, if a node is determined to reach the target state safely without requiring further control inputs, the search process is successfully terminated. If all feasible nodes have been explored and no valid path is found, the algorithm concludes with a failure to identify a feasible solution.

Algorithm 1: Graph-Based Collision Avoidance Planning Method

```
Initialization:
1. Create a node n_start from s_start, set n_start.g = 0.
2. Compute n_start.f = Priority(n_start, s_goal).
3. Insert n_start into the open set O (a priority queue keyed by f-value).
4. Let the closed set C = ∅.
while O is not empty do
    Extract n_curr from O with the smallest f-value;
    if NoCollision(FreeMotion(n_curr), 𝒪) and FinalState(n_curr) ∈ 𝒳 then
        return GeneratePath (n_curr);
    Add n_curr to C;
    foreach Maneuver decision u ∈ 𝒰 do
        n_new ← CalcNextState(n_curr, u);
        if n_new is not in C then
            if CheckCollision (n_new, 𝒪) = False then
                g_temp ← n_curr.g + ||u||_2;
                if n_new not in O or g_temp < n_new.g then
                    n_new.g ← g_temp;
                    n_new.f ← Priority(n_new, s_goal);
                    n_new.parent ← n_curr;
                    if n_new not in O then
                        Insert n_new into O;
                    else
                        Update n_new's priority in O;
return No Feasible Path Found;
```

3.2.1. Collision Check Conjunctions between spacecraft and debris occur instantaneously, with relative velocities reaching as high as 10 km/s. During this extremely short encounter period, the distance between the two objects changes rapidly, and collision risk is generally assessed based on the minimum separation distance at TCA. The obstacle space associated with $Debris_n$ is defined as follows:

$$\mathcal{O}_n = \left\{ \boldsymbol{x}(t) \;\middle|\; t = t^n_{\mathrm{TCA}},\ \Delta r^n_{\min} = \left\| \left(I_3 - e_n e_n^\top \right) \tilde{\boldsymbol{r}}_n \left(t^n_{\mathrm{TCA}} \right) \right\| < D_{\mathrm{threshold}} \right\} \tag{4}$$

where $e_n = \tilde{\boldsymbol{v}}_n \left(t^n_{\mathrm{TCA}} \right) / \| \tilde{\boldsymbol{v}}_n \left(t^n_{\mathrm{TCA}} \right) \|$ is the unit vector of the relative velocity of $Debris_n$ with respect to the spacecraft. $\tilde{\boldsymbol{r}}_n \left(t^n_{\mathrm{TCA}} \right)$ denotes the relative position vector of debris n with respect to the spacecraft at TCA_n, and $\tilde{\boldsymbol{v}}_n \left(t^n_{\mathrm{TCA}} \right)$ represents the corresponding relative velocity vector at the same instant. I_3 is the 3-dimensional identity matrix. $D_{threshold}$ is the predefined safety distance. The overall obstacle space, composed of all debris objects, is expressed as the union of individual debris obstacle space:

$$\mathcal{O} = \bigcup_{n=1}^{N} \mathcal{O}_n \tag{5}$$

We further define the station-keeping constraint to ensure that the spacecraft remains within an acceptable deviation envelope relative to its nominal orbit throughout the mission duration. Specifically, this constraint imposes both position and velocity bounds in the RTN frame at every time instant t. The admissible configuration space $\mathcal{X}$ is therefore defined as:

$$\mathcal{X} = \left\{ \boldsymbol{x}(t) = \begin{bmatrix} \boldsymbol{r}_{\mathrm{RTN}}(t) \\ \boldsymbol{v}_{\mathrm{RTN}}(t) \end{bmatrix} \;\middle|\; \boldsymbol{r}_{\min} \leq \boldsymbol{r}_{\mathrm{RTN}}(t) \leq \boldsymbol{r}_{\max},\ \boldsymbol{v}_{\min} \leq \boldsymbol{v}_{\mathrm{RTN}}(t) \leq \boldsymbol{v}_{\max} \right\} \tag{6}$$

where $\boldsymbol{r}_{\mathrm{RTN}}(t)$ is the spacecraft's position vector in the RTN frame at time t, $\boldsymbol{v}_{\mathrm{RTN}}(t)$ is the corresponding velocity vector, $\boldsymbol{r}_{\min}$, $\boldsymbol{r}_{\max}$ are the lower and upper bounds on the allowable position deviation, $\boldsymbol{v}_{\min}$, $\boldsymbol{v}_{\max}$ are the lower and upper bounds on the allowable velocity deviation.

To ensure compliance with the station-keeping configuration constraints, the planned trajectory must satisfy:

$$\boldsymbol{x}(t) \in \mathcal{X}, \quad \forall t \in [t_0, t_f] \tag{7}$$

where t_0 and t_f denote the initial and final times of the trajectory planning process, respectively. Violation of this constraint would render the corresponding trajectory infeasible, as it could result in mission degradation or interference with other operational systems.

3.2.2. Analytic Expansion To further improve the search efficiency, an analytic expansion is performed at each explored node. Utilizing the Clohessy-Wiltshire state transition equations, the future states of the spacecraft at each debris conjunction time are rapidly computed under the assumption of no further control inputs. These predicted states are then evaluated using the collision detection function to determine potential collision risks. Additionally, the state of the spacecraft at the final target time is calculated, and its deviation from the desired goal state is assessed.

If, under zero-control propagation, the current node is free of collisions with all upcoming debris objects and the deviation from the goal state at the final time satisfies the required configuration constraints, the search process is terminated, and the current trajectory is accepted as a feasible solution. In this case, no further maneuvering is necessary, as a collision-free and goal-compliant path has been identified.

3.2.3. Priority Function Design In the proposed graph search algorithm, the design of the priority function determines which nodes in the queue are expanded first. An effective priority function helps to avoid exploring excessive numbers of non-promising nodes, thereby directly improving the computational efficiency of the algorithm. At the same time, the node expansion order governed by the priority function also affects the optimality of the resulting solution. In this study, we design three distinct priority strategies: (1) a Dijkstra-like strategy that prioritizes nodes based on accumulated fuel cost; (2) a depth-first strategy that prioritizes nodes by their temporal depth; and (3) an A*-like strategy that incorporates a heuristic estimate of the future collision avoidance cost.

Specifically, Strategy 1 aims to optimize fuel consumption; Strategy 2 prioritizes faster search speed; while Strategy 3 offers a balanced trade-off between computational efficiency and solution optimality. The following sections describe the implementation of each strategy in detail.

(1) Fuel-Optimal Strategy

$$f_{priority} = Node.cost \tag{8}$$

Our approach is Dijkstra-like rather than an exact implementation, as nodes represent continuous state regions, and the goal is defined as a region, not a point. As grid resolution

Journal of Physics: Conference Series **3056** (2025) 012006 doi:10.1088/1742-6596/3056/1/012006

increases, the algorithm asymptotically approximates Dijkstra's behavior, though at the cost of higher computational effort due to the expanded search space.

(2) Time-Depth Priority Strategy

$$f_{priority} = -Node.time \tag{9}$$

In the time-depth priority strategy, the algorithm prioritizes nodes at later time steps, effectively adopting a depth-first approach. This narrows the early search space and accelerates exploration into deeper solution regions. In multi-debris scenarios, deeper layers often represent trajectories that have avoided more threats, increasing feasibility. By focusing computational resources on these nodes, the strategy improves search efficiency, particularly in complex collision environments.

(3) Heuristic-Guided Search Strategy

The followings introduce a heuristic-guided search strategy to balance optimality and efficiency:

$$f_{priority} = \text{Node.cost} + h(\text{Node}) \tag{10}$$

where $h(\text{Node})$ is a heuristic function designed to estimate the fuel cost required for future debris avoidance and configuration recovery.

To efficiently approximate future avoidance cost, we adopt a directional decomposition strategy. Specifically, candidate maneuver directions are limited to the principal axes of the RTN frame, corresponding to the radial (R), along-track (T), and cross-track (N) directions. For each candidate direction, the required impulse magnitude is analytically computed by solving a scalar quadratic equation, and the minimum value among these candidates is selected as the heuristic estimate.

For a given candidate direction $\boldsymbol{d} \in \{\boldsymbol{e}_x, \boldsymbol{e}_y, \boldsymbol{e}_z\}$, where $\boldsymbol{e}_x = [1, 0, 0]^\top$, $\boldsymbol{e}_y = [0, 1, 0]^\top$, and $\boldsymbol{e}_z = [0, 0, 1]^\top$, the avoidance maneuver is assumed as:

$$\Delta \boldsymbol{V} = \alpha \boldsymbol{d} \tag{11}$$

Using the Clohessy-Wiltshire (CW) equations, the spacecraft's relative position at the time of conjunction can be expressed as:

$$\boldsymbol{r}_{\text{rel}} = \boldsymbol{r}_{\text{rel},0} - \boldsymbol{\Phi}_{rv} \Delta \boldsymbol{V} \tag{12}$$

According to Eq.4, the collision constraint is:

$$\left\| \left(\boldsymbol{I} - \boldsymbol{e}\boldsymbol{e}^\top \right) \boldsymbol{r}_{\text{rel}} \right\| \geq D_{\text{threshold}} \tag{13}$$

By substituting Eq.11 and Eq.12 into Eq.13, and noting that the optimal maneuver corresponds to the case where the equality holds, the following equation is obtained:

$$A\alpha^2 + B\alpha + C = 0 \tag{14}$$

where $A = \boldsymbol{d}^\top \Phi_{rv}^\top \boldsymbol{P} \Phi_{rv} \boldsymbol{d}$, $B = -2\boldsymbol{r}_{\text{rel},0}^\top \boldsymbol{P} \Phi_{rv} \boldsymbol{d}$, $C = \boldsymbol{r}_{\text{rel},0}^\top \boldsymbol{P} \boldsymbol{r}_{\text{rel},0} - D_{\text{threshold}}^2$ and $\boldsymbol{P} = \boldsymbol{I} - \boldsymbol{e}\boldsymbol{e}^\top$.

The minimum required impulse magnitude along direction $\boldsymbol{d}$ is given by:

$$\alpha^* = \frac{-B + \sqrt{B^2 - 4AC}}{2A} \tag{15}$$

provided that the discriminant satisfies $B^2 - 4AC \geq 0$; otherwise, the direction is considered infeasible.

Journal of Physics: Conference Series **3056** (2025) 012006 doi:10.1088/1742-6596/3056/1/012006

Finally, the directional strategy selects the maneuver with the minimal impulse magnitude among all evaluated directions:

$$\Delta \boldsymbol{V}^* = \alpha^*_{\min} \cdot \boldsymbol{d}_{\min} \tag{16}$$

where:

$$\boldsymbol{d}_{\min} = \arg \min_{\boldsymbol{d} \in \{\boldsymbol{e}_x, \boldsymbol{e}_y, \boldsymbol{e}_z\}} \alpha^*(\boldsymbol{d}) \tag{17}$$

If no future collision risks are detected, the heuristic function estimates the fuel cost required to transfer the spacecraft to the goal state. Assuming a two-impulse maneuver, executed at the current time and at the final target time, the estimated fuel consumption is given by:

$$\Delta V_1 = \phi_{rv}^{-1} \left(\boldsymbol{r}_{goal} - \phi_{rr} \boldsymbol{r} \right) \tag{18}$$

$$\Delta V_2 = \phi_{vr} \boldsymbol{r} + \phi_{vv} \left(\boldsymbol{v} + \Delta V_1 \right) - \boldsymbol{v}_{goal} \tag{19}$$

Thus, the total estimated cost is:

$$h(\text{Node}) = \Delta V = \|\boldsymbol{\Delta V_1}\| + \|\boldsymbol{\Delta V_2}\| \tag{20}$$

For general A* algorithm if the heuristic function $h(n)$ consistently underestimates or accurately estimates the true remaining cost then the algorithm is guaranteed to find the optimal solution. In the heuristic design adopted in this study, the avoidance cost estimation considers only individual debris objects, which naturally yields a lower estimate than the total cost of avoiding multiple debris simultaneously. Furthermore, the configuration recovery cost is computed under the assumption of an ideal two-impulse maneuver, without accounting for thrust limitations, timing constraints, or potential collision risks. Consequently, the heuristic function provides an admissible cost estimate, ensuring that the fuel-optimality of the resulting solution is preserved.

4. Simulation Validation

In this section, simulation experiments are conducted to validate the effectiveness of the proposed algorithm in planning collision avoidance trajectories for space debris scenarios. Three representative cases are designed: a single-debris scenario, a two-debris scenario, and a multiple-debris scenario.

4.1. Single-Debris Scenario

This subsection verifies the proposed algorithm's effectiveness in a single-debris avoidance scenario. The satellite operates in a near-circular orbit (semi-major axis: 7000 km, zero inclination), while the debris is placed in a polar orbit, creating a 0.1-km close approach at TCA. A 2 km minimum safe distance and strict orbit-keeping constraints ($\leq 2\,\text{km}$ position deviation, $\leq 0.1\,\text{m/s}$ velocity deviation) are enforced.

The control interval is set to 1200 seconds over an 18,000-second simulation. The satellite performs impulsive maneuvers in radial and along-track directions, with a maximum impulse of $2.5 \times 10^{-5}\,\text{km/s}$. As no out-of-plane maneuvers are applied, the spacecraft's motion is confined to the O-XY plane. The debris safety zone is projected onto the O-XY plane, forming an elliptical region (as illustrated in Fig. 2).

The avoidance trajectories generated by the different methods are shown in Fig.4, with key performance metrics summarized in Table.1. The black dashed line indicates the debris safety boundary, and the black diamonds mark the spacecraft's position at the TCA. All strategies achieve safe avoidance and respect configuration constraints. The time-depth priority strategy is fastest (1.34 s, 1,031 nodes) but consumes the most fuel. The fuel-optimal strategy minimizes

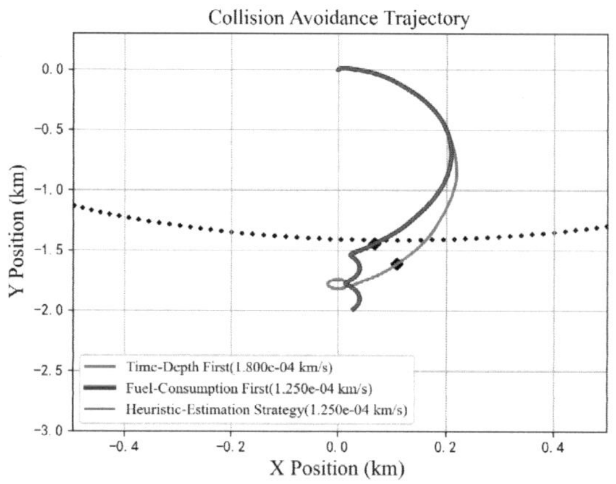

Figure 4: Collision Avoidance Trajectory in the Single-Debris Scenario.

Table 1: Planning Performance by Different Methods in the Single-Debris Scenario.

Method	**Expanded Node Number**	**Time Cost**(s)	**Fuel Cost**(km/s)
Time-Depth First	1,031	1.34	1.80×10^{-4}
Fuel-Consumption First	15,322	15.613	1.25×10^{-4}
Heuristic-Estimation Strategy	1,972	5.27	1.25×10^{-4}

consumption but requires longer computation (15.61 s). The heuristic-guided strategy achieves near-optimal fuel efficiency with reduced effort (5.27 s, 1,972 nodes), offering the best overall balance.

4.2. Two-Debris Scenario

Table 2: Collision Avoidance Trajectory in the Two-Debris Scenario

Method	**Expanded Node Number**	**Time Cost**(s)	**Fuel Cost**(km/s)
Time-Depth First	2,505	7.23	2.85×10^{-4}
Fuel-Consumption First	23,322	89.54	1.55×10^{-4}
Heuristic-Estimation Strategy	15,672	60.12	1.55×10^{-4}

In the two-debris conjunction scenario, the initial optimal maneuver to avoid Debris 1 would lead the spacecraft into Debris 2's collision region. To resolve this, the spacecraft adjusts its trajectory through the narrow corridor between the two threats. This case reflects an extreme situation in dense debris environments. Fig.5 and Table.2 summarize the trajectories and performance comparisons.

All methods successfully avoided both debris while respecting configuration constraints. However, strategy differences became more evident. The time-depth priority strategy delivered the fastest result (7.23 s, 2,505 nodes) but at higher fuel consumption $\left(2.85 \times 10^{-4}\,\text{km/s}\right)$. The fuel-optimal strategy achieved the lowest fuel use $\left(1.55 \times 10^{-4}\,\text{km/s}\right)$, but at high computational cost (23,322 nodes, 90 s). In contrast, the heuristic-guided strategy matched the optimal fuel performance while significantly reducing computation time (60.12 s) and node expansions

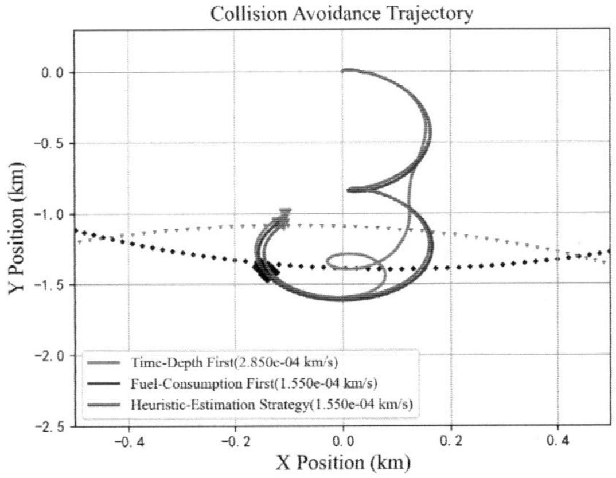

Figure 5: Planning Performance Across Different Methods in the Two-Debris Scenario.

(15,672). These results confirm the heuristic-guided strategy's advantage in balancing efficiency and solution quality for complex, multi-conjunction avoidance tasks.

4.3. Multiple-Debris Scenario

In this section, we conduct simulations to validate the capability of the proposed algorithm in planning safe trajectories for multiple debris encounters. In the scenario design, we randomly generate 10 debris objects that will have close approaches with the satellite. These debris are parameterized by randomly generating conjunction time, velocity angle and conjunction distance vector. All debris are set to have a conjunction distance of less than 1 km. The complex avoidance trajectory planned by the spacecraft in this high-risk environment is illustrated in Fig. 7, and the performance comparison of the different methods is summarized in Table 3.

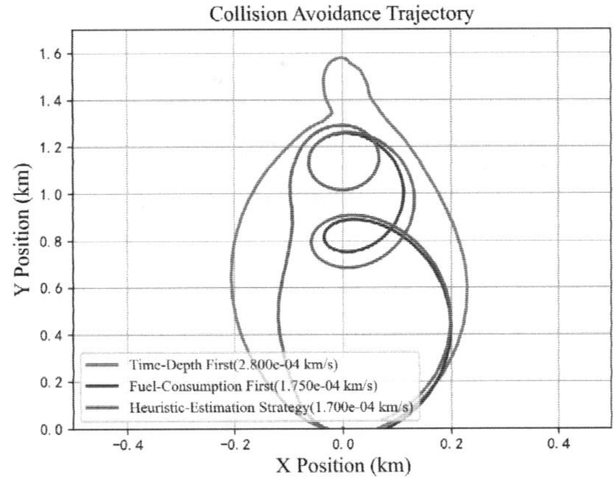

Figure 6: Collision Avoidance Trajectory in the Multi-Debris Scenario.

Simulation results show that all three algorithms successfully avoided multiple debris threats, keeping the spacecraft outside all safety boundaries and verifying the robustness of the proposed methods. The time-depth priority strategy achieved the fastest search (9.81 s), but with higher fuel consumption $(2.80 \times 10^{-4}\,\text{km/s})$, making it suitable for real-time scenarios with moderate

Table 3: Planning Performance Across Different Methods in the Multi-Debris Scenario

Method	**Expanded Node Number**	**Time Cost**(s)	**Fuel Cost**(km/s)
Time-Depth First	2,850	9.81	2.89×10^{-4}
Fuel-Consumption First	34,459	140.23	1.75×10^{-4}
Heuristic-Estimation Strategy	22,387	120.38	1.70×10^{-4}

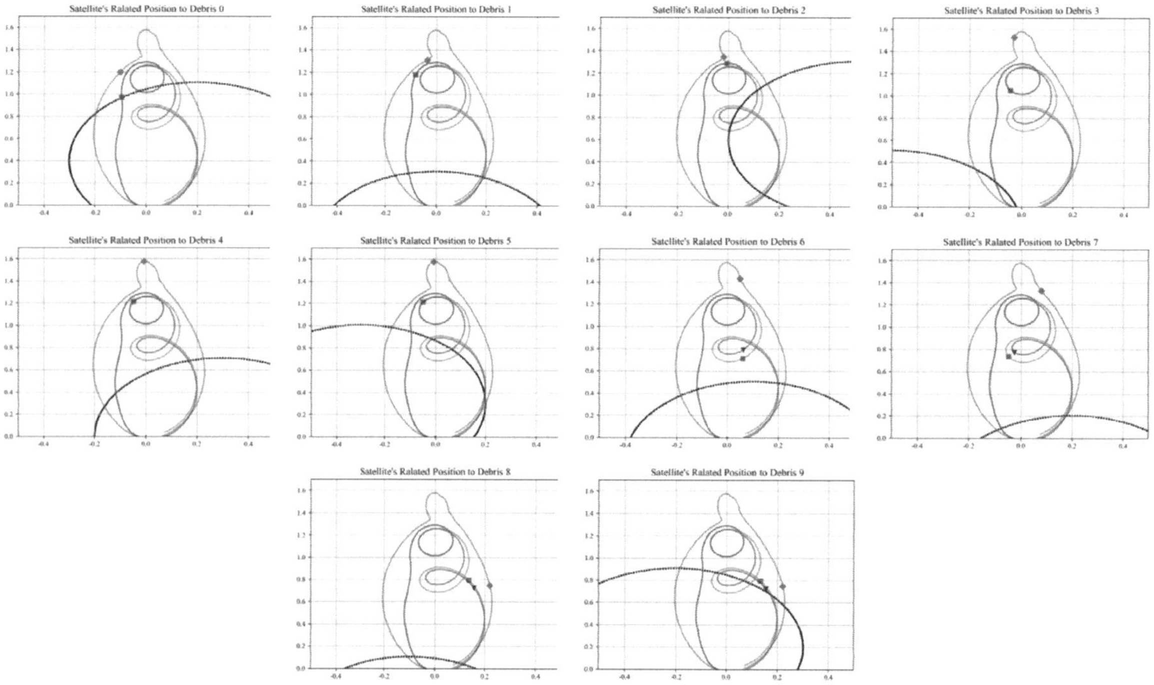

Figure 7: Satellite Relative Positions at Conjunctions with Ten Debris Objects.

accuracy demands. The fuel-optimal strategy minimized fuel use $(1.75 \times 10^{-4}\,\text{km/s})$ but at high computational cost (34,459 nodes, 140.23 s). In contrast, the heuristic-guided strategy offered the best balance, achieving the lowest fuel consumption $(1.70 \times 10^{-4}\,\text{km/s})$ while significantly reducing computation time (120.38 s) and node expansions (22,387). These results confirm the scalability and practical advantage of the heuristic-guided approach for complex, multi-target avoidance tasks.

4.4. Further Discussion

Although both the fuel-optimal and heuristic-guided strategies achieved near-identical total fuel consumption close to the global optimum, their resulting trajectories still differ, as shown in Fig.5 and Fig.6. These differences arise not from algorithmic errors but from inherent characteristics of the planning methods, primarily due to three factors. First, the search space employs continuous node representation and a soft-constrained terminal region, where nodes correspond to feasible state regions rather than fixed points, and the terminal condition allows multiple equivalent solutions. This design avoids abrupt cost changes but introduces solution diversity near the goal. Second, the algorithms use different priority functions, shaping their search behaviors. The fuel-optimal strategy exhaustively explores lower-cost paths, while the heuristic-guided strategy incorporates predictive cost estimates, focusing exploration on promising regions. These differences lead to variations in maneuver timing and thrust direction.

In essence, these trajectory variations reflect the natural flexibility of the algorithmic framework rather than suboptimality. Both methods reliably produce feasible, near-optimal solutions, demonstrating the robustness and adaptability of the approach in complex multi-threat scenarios.

5. Conclusion

With the rapid increase of space debris, satellites face growing collision risks that traditional strategies, focused on isolated events, fail to address effectively in multi-conjunction scenarios. To tackle this challenge, this study proposes a graph-based trajectory planning framework for systematic avoidance in complex orbital environments. The framework discretizes the spacecraft's state space by explicitly introducing time and employs a continuous node representation to enhance search flexibility. A rigorous obstacle space model is constructed, defining collision criteria based on relative motion at conjunction, alongside configuration constraints to maintain orbital safety. An analytical expansion strategy accelerates the search by efficiently pruning infeasible nodes. Furthermore, three priority-based search strategies are developed: a fuel-optimal approach, a time-depth priority strategy favoring rapid solutions, and a heuristic-guided method that balances efficiency and path quality using predictive cost estimation. Simulations in different scenarios demonstrate the framework's ability to generate feasible, constraint-compliant trajectories. Notably, the heuristic-guided strategy achieves near-optimal fuel consumption while substantially reducing computational effort compared to the fuel-optimal method. Although variations exist among strategies due to node continuity and goal region flexibility, all maintain robust, high-quality solutions. In summary, the proposed method provides an effective solution for autonomous and safe navigation of spacecraft in densely populated space debris environments. Future work will focus on incorporating higher-fidelity dynamical models and extending the approach to scenarios involving continuous-thrust maneuvers.

References

[1] McDowell J C 2020 *The Astrophysical Journal Letters* **892** L36
[2] Kessler D J, Johnson N L, Liou J and Matney M 2010 *Advances in the Astronautical Sciences* **137** 2010
[3] Bombardelli C and Hernando-Ayuso J 2015 *Journal of Guidance, Control, and Dynamics* **38** 217–225
[4] Gonzalo J L, Colombo C and Di Lizia P 2021 *Journal of Guidance, Control, and Dynamics* **44** 469–487
[5] Morselli A, Armellin R, Di Lizia P and Bernelli-Zazzera F 2014 *Advances in the Astronautical Sciences* **152** 1819–1838
[6] Ströbel R and Stoll E 2024 *Journal of Guidance, Control, and Dynamics* **47** 358–365
[7] Hernando-Ayuso J and Bombardelli C 2021 *Journal of Guidance, Control, and Dynamics* **44** 983–995
[8] Kim E H, Kim H D and Kim H J 2012 *Journal of Astronomy and Space Sciences* **29** 11–21
[9] Nag S, Murakami D D, Marker N A, Lifson M T and Kopardekar P H 2021 *Acta Astronautica* **180** 489–506
[10] Masson M, Arzelier D, Joldes M, Revelin B and Thomassin J 2023 *IFAC-PapersOnLine* **56** 1995–2000
[11] Pavanello Z, Pirovano L, Armellin R, De Vittori A and Di Lizia P 2024 *AIAA Scitech 2024 Forum* p 0845
[12] Pavanello Z, Pirovano L and Armellin R 2024 *Journal of Guidance, Control, and Dynamics* **47** 1855–1871
[13] Dolgov D, Thrun S, Montemerlo M and Diebel J 2008 *ann arbor* **1001** 18–80
[14] Chen J, Li M, Yuan Z and Gu Q 2020 *2020 IEEE 4th Information Technology, Networking, Electronic and Automation Control Conference (ITNEC)* vol 1 (IEEE) pp 958–962

Journal of Physics: Conference Series **3056** (2025) 012007 doi:10.1088/1742-6596/3056/1/012007

Combined Euler-Lagrange Method for 3D Large Droplet Collection Efficiency Prediction

Jiaxin Yu[1], Weijian Chen[1*][0000-0001-6776-4492], Long Xia[1], Ning Guan[2]

[1] Key Laboratory of Aircraft Environment Control and Life Support, MIIT, Nanjing University of Aeronautics & Astronautics, Nanjing, China, 210016

[2] College of Aeronautics, Shandong Jiaotong University, No.5001, Haitang Road, Changqing University Science and Technology Park, Jinan, China, 250037

*Corresponding author's e-mail: weijian_chen@nuaa.edu.cn

Abstract. In this paper, we propose a combined Euler-Lagrange method for calculating large three-dimensional water droplets. We use the Euler method to determine the initial droplet collection efficiency and then write UDF code to process the splash and obtain the collection efficiency after splashing. The re-entry calculation for the droplets is done using the DPM approach. We utilize a three-dimensional model of the DLR-F6 wing-body combination to test the effectiveness of this method, while also comparing results calculated using FENSAP. The results show that the splashing results processed by FENSAP are slightly more conservative than those in this paper. The DPM re-entry calculation results indicate that the wing root region captures additional droplets due to droplet re-entry, demonstrating that the DPM code presented in this paper can be applied for re-entry calculations of large droplets in three-dimensional scenarios.

1. INTRODUCTION

Supercooled Large Droplets (SLD) refer to water droplets in cloud layers with a Mean Volumetric Diameter (MVD) exceeding 50 μm. Distinct from conventional small droplets, SLDs exhibit complex dynamic behaviors during motion, including deformation, breakup, and splashing, posing significant challenges to the prediction of droplet impingement characteristics.

In recent years, extensive research on large droplets has been conducted by domestic and international scholars. Honsek et al. [1,2]transformed the original semi-empirical splash model to develop a mathematical formulation compatible with the Eulerian method. Iuliano et al.[3] defined mass and momentum source terms on airfoil wall cells to integrate SLD dynamics models (including splashing and rebound) into an Eulerian droplet impingement computational framework. David R. Bilodeau et al. [4]proposed a conservative Eulerian numerical approach for simulating SLD splashing and rebound on aircraft surfaces, which incorporates the effects of post-impact droplets (secondary impingement) through sequential solutions of the conservation equations.

C. Wang et al.[5,6] employed the Discrete Phase Model (DPM) to track droplet trajectories and proposed a two-dimensional splash model to investigate the dynamic characteristics of SLD impingement on walls. Xiaobin Shen et al.[7] developed a Lagrangian approach to study the influence of upstream components and air injection on the impingement characteristics of downstream airfoil surfaces. Han Han et al.[8] introduced a hybrid sampling algorithm based on the Rosin-Rammler distribution function to simulate SLD particle size distribution (PSD), achieving highly accurate numerical icing simulations with realistic SLD droplet size spectra.

Bellosta et al. [9] proposed a Lagrangian particle-tracking framework for water droplet collection efficiency calculation, incorporating an automatic resolution adaptation method and describing a parallel computation strategy. Within the Eulerian framework, they introduced a relaxation-based numerical scheme for solving the hyperbolic system of droplet transport equations. Furthermore, the authors presented a hybrid Eulerian-Lagrangian approach specifically designed for collection efficiency computation under supercooled large droplet (SLD) conditions, which enables quantitative evaluation of secondary droplet contributions from splashing effects.

Despite extensive research by numerous scholars on Supercooled Large Droplets (SLD), the coupled dynamic characteristics including deformation, breakup, and splashing significantly complicate the prediction of wall surface water collection efficiency, and current understanding of SLD remains incomplete. This paper proposes a coupled Eulerian-Lagrangian methodology for three-dimensional water collection rate prediction of large droplets. The approach involves: (1) developing an Eulerian droplet impingement computational program with customized code implementation for SLD splashing post-processing to determine collection efficiency on impacted surfaces; (2) employing the Discrete Phase Model (DPM) to simulate water droplet re-entrainment by releasing Lagrangian particles from impingement surfaces, thereby quantifying their contribution to overall surface collection efficiency.

2. Calculation method for droplet collection coefficient

The calculation methods for water collection efficiency mainly include the Eulerian approach and Lagrangian approach. In the Eulerian method, water droplets are treated as a continuous phase by introducing volume fraction to establish mass and momentum conservation equations for the droplet phase, where the collection efficiency is obtained by solving the droplet volume fraction and velocity variables across all grid cells in the droplet field. The Lagrangian method establishes the equations of motion for individual droplet particles and calculates surface impingement characteristics by solving particle trajectories. Since the collection efficiency at any surface location in the Lagrangian method needs to be computed based on surrounding impact point positions and initial release point positions, the calculation becomes particularly complex for three-dimensional configurations. While the Eulerian method offers relatively simpler collection efficiency computation, it faces difficulties in accurately describing large droplet motion processes and may encounter convergence issues. The following describes these two methods involved in this study.

2.1. Euler's Method

The Eulerian approach for water droplet simulation introduces the droplet volume fraction α, defined as the ratio of droplet volume to control volume within a computational cell. Given the sparse distribution of atmospheric droplets where α typically reaches magnitudes of 10^{-6}, the influence of droplets on airflow is generally neglected, allowing the two-phase (air-droplet) system to be simplified as a one-way coupled single-phase solution. For small droplets (MVD < 50 μm), the steady-state differential forms of the continuity equation and momentum equation governing droplet motion are expressed as:

$$\nabla \cdot \left(\boldsymbol{u}_w \alpha_w\right) = 0 \tag{1}$$

$$\nabla \cdot \left(\alpha_w \boldsymbol{u}_w \boldsymbol{u}_w\right) = \alpha_w K \left(\boldsymbol{u}_a - \boldsymbol{u}_w\right) + \alpha_w \boldsymbol{F} \tag{2}$$

In the formulation, α_w represents the droplet volume fraction, $\boldsymbol{u}_w$ denotes the droplet velocity vector, $\boldsymbol{u}_a$ is the air velocity vector, and $\boldsymbol{F}$ accounts for additional body forces beyond drag. The momentum exchange coefficient K is expressed as:

$$K = \frac{3\mu_a C_D Re_r}{4\rho_w d_w^2} \tag{3}$$

In the formulation, d_w μ_a represents the aerodynamic viscosity, $\boldsymbol{u}_w$ denotes the droplet diameter, and ρ_w is the droplet density. The drag coefficient C_D is defined as a function of the relative Reynolds number Re_r, expressed as:

$$Re_r = \frac{\rho_a \| \boldsymbol{u}_a - \boldsymbol{u}_w \| d_w}{\mu_a} \tag{4}$$

Generally, both sides of Equations (1) and (2) are normalized by dividing by the freestream volume fraction α_∞ to achieve dimensionless form α_w .Here, the value of α_∞ is set to LWC_∞ / ρ_w, and the far-field boundary condition for volume fraction is uniformly specified as 1. This approach facilitates code implementation in numerical simulations.

2.2. Lagrangian DPM method

The following outlines our implementation process using the Discrete Phase Model (DPM) module in ANSYS Fluent with UDF for two-dimensional large droplet simulation and water collection efficiency calculation:

$$\frac{d\boldsymbol{u}_w}{dt} = K(\boldsymbol{u}_a - \boldsymbol{u}_w) \tag{5}$$

The expression for K has been provided in Equation (3). For large droplet conditions, the drag coefficient C_D in K is specified programmatically via the DEFINE_DPM_DRAG macro. Droplet splashing is implemented through the DEFINE_DPM_BC macro, which determines whether a droplet particle deposits onto the airfoil surface or re-enters the flow field based on the splashing threshold criterion established in the splashing model.Given that splashing involves mass loss, the re-entrained droplets experience progressive mass attenuation. To quantify this effect, a DPM scalar variable is defined using the DEFINE_DPM_SCALAR_UPDATE macro to track the mass reduction coefficient g_n after the n-th splashing event, formulated as:

$$g_n = f_n g_{n-1} \tag{6}$$

f_n denotes the splashing mass loss ratio, representing the mass ratio between re-entrained droplets after the n-th splash event and the initial incident droplet mass, initialized as $g_0=1$. Assuming a sufficient number N of Lagrangian mass particles are released with uniform spacing along a vertical span of length Δy in the undisturbed far-field region, the contribution to local collection efficiency β_i at the i-th airfoil surface element with area ΔS_i from an impinging particle is given by:

$$\beta_i = \beta_i + (1 - f_n) g_{n-1} \frac{\Delta y}{N \Delta S_i} \tag{7}$$

When all particle trajectories have been tracked, the cumulative calculation of wall surface collection efficiency is completed.

To validate the reliability of the aforementioned 2D DPM large droplet (SLD) UDF code, verification cases were conducted based on droplet impingement characteristic experiments performed by Papadakis et al. in the NASA IRT wind tunnel using a smooth NACA 23012 airfoil.

The test airfoil featured a standard chord length of 0.914 m and was evaluated under controlled wind tunnel conditions with a freestream air velocity of 78.25 m/s, an angle of attack of 2.5°, and an incoming static pressure of 94.803 kPa. The droplet characteristics followed FAR 25 Appendix O specifications, with two distinct test cases: one using droplets with a MVD of 111 μm paired with a LWC of 0.73 g/m³, and another employing SLD of 236 μm MVD with 1.89 g/m³ LWC to represent severe icing conditions.

Through grid independence verification, the final computational mesh was determined to consist of 100,000 cells.

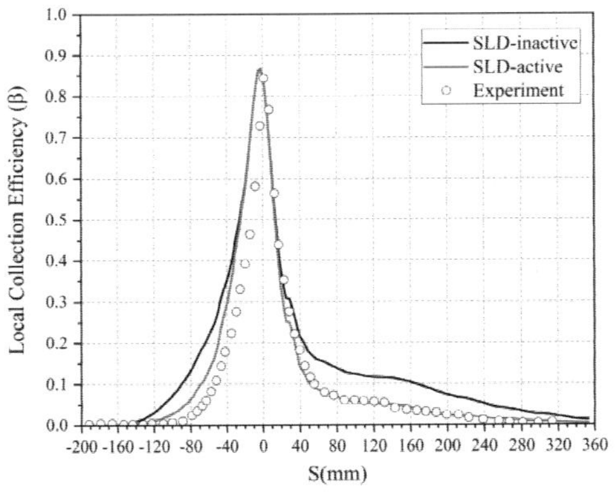

Figure 1. Local Water Collection Efficiency Profile, MVD=111μm

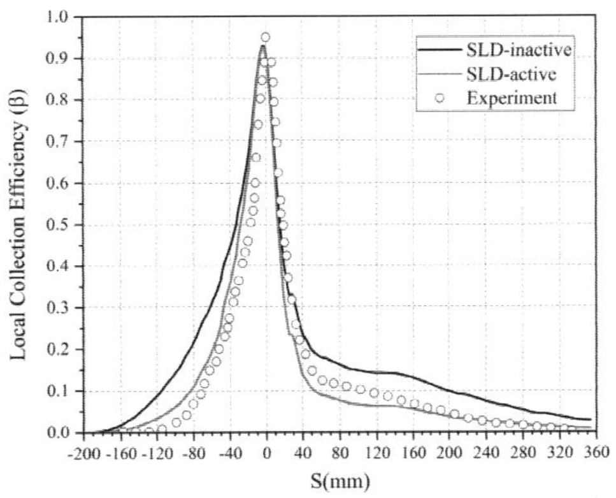

Figure 2. Local Water Collection Efficiency Profile, MVD=236μm

After computing 20 datasets, the integrated collection efficiency was compared against experimental results. Figures 1 and 2 present comparative analyses of calculated droplet collection efficiency with and without SLD model activation versus experimental data, where the abscissa S denotes the curvilinear coordinate along the airfoil surface. The results demonstrate significant discrepancies between conventional (non-SLD) model predictions and experimental measurements for large droplets. In contrast, the SLD model incorporating droplet deformation, breakup, rebound, and splashing phenomena exhibits markedly improved agreement with test data, thereby validating the reliability of the present computational methodology.

3. Splash and re-entry calculations of 3D models

For three-dimensional complex surfaces, it is necessary to release and track a sufficient number of particles to capture droplet impingement zones and areas with high surface curvature. This approach demands substantial computational memory and time, making the DPM method difficult to apply. Instead, a Eulerian-based UDF program is employed to simulate large droplet impingement on 3D surfaces and obtain the primary droplet collection efficiency. Subsequently, a post-processing approach is used to account for collection efficiency loss during the splashing process. For droplet re-entrainment, a novel DPM-based re-entrainment calculation method is proposed, achieving a combined computation integrating both approaches.

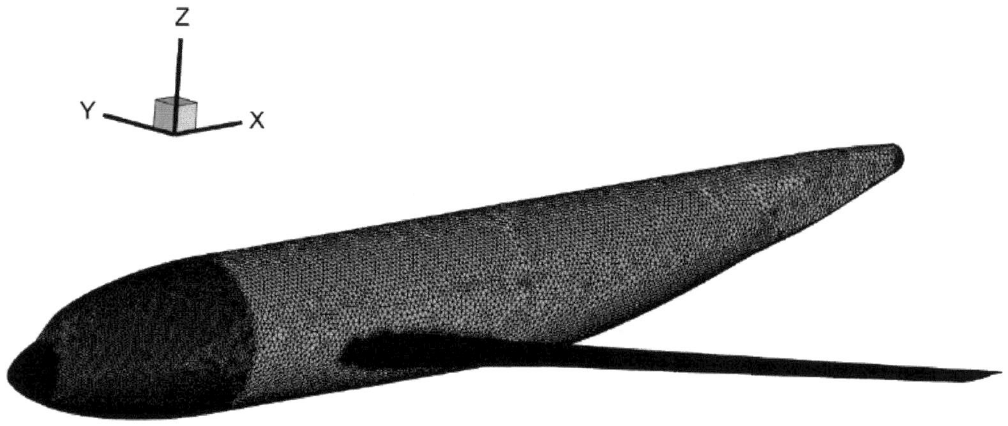

Figure 3. Schematic diagram of the wing body surface grid

Table 1. Three-dimensional large water droplet calculation conditions

Calculated parameters	Valid values
Mach number	0.3
Static temperature(K)	268
Flight altitude(m)	2000
Angle of attack	0
LWC(g/m^3)	0.5
MVD(μm)	236

The computational demonstration employs a three-dimensional DLR-F6 wing-body configuration model, sourced from the official website of the 2nd Drag Prediction Workshop.[10] This benchmark model features a mean aerodynamic chord (MAC) of 141.2 mm. The computational analysis selected droplets with a Median Volume Diameter (MVD) of 236 μm.The computational mesh was generated using ICEM CFD software with all-tetrahedral unstructured elements, incorporating local refinement at critical impingement zones including the wing leading edge and nose section. Figure 3 illustrates the surface mesh distribution on the wing-body assembly, comprising 4.33 million volume cells and 1.26 million grid nodes. The simulation conditions are specified in Table 1.

Figure 4 displays the primary water collection efficiency distribution on the wing-body surface computed via UDF, while Figure 5 presents the post-processed collection efficiency after accounting for splashing effects. Comparative analysis reveals that the post-processed results exhibit reduced collection efficiency values due to the mass subtraction of splashed droplets departing from the surface, demonstrating the physical consistency of the implemented methodology.

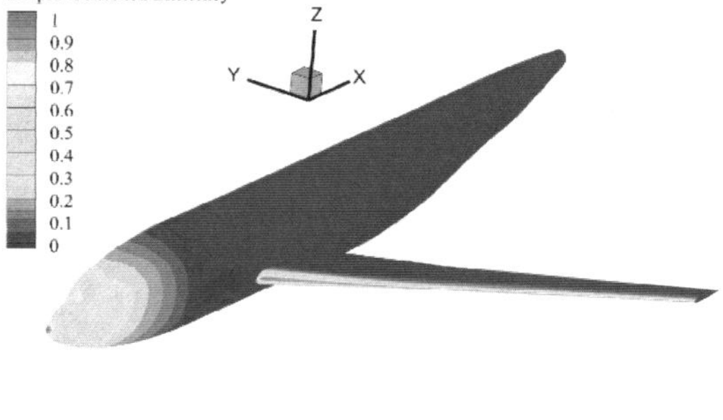

Figure 4. The initial collection rate of water droplets calculated by UDF.

Figure 5. Post-processing splash droplet collection rate calculated by UDF

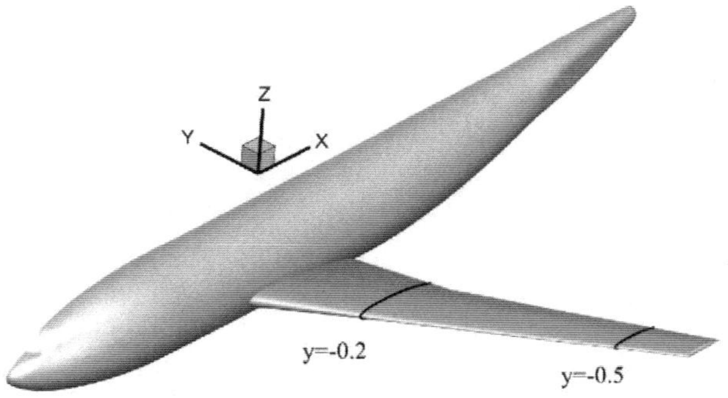

Figure 6. Schematic diagram of slice location（y=-0.2m，y=-0.5m）

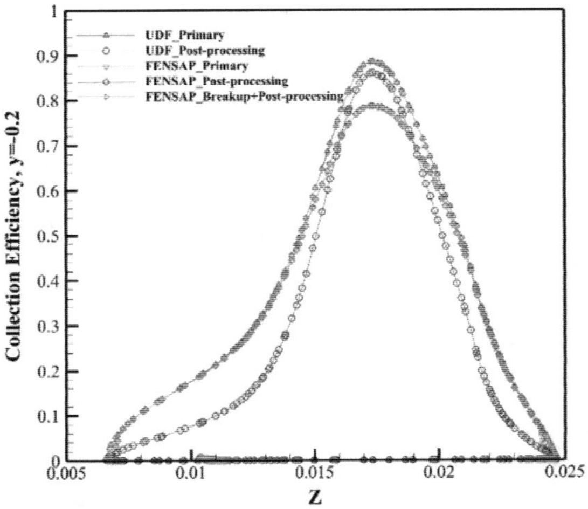

Figure 7. Comparison curve of collection rates at the section.(y=-0.2m)

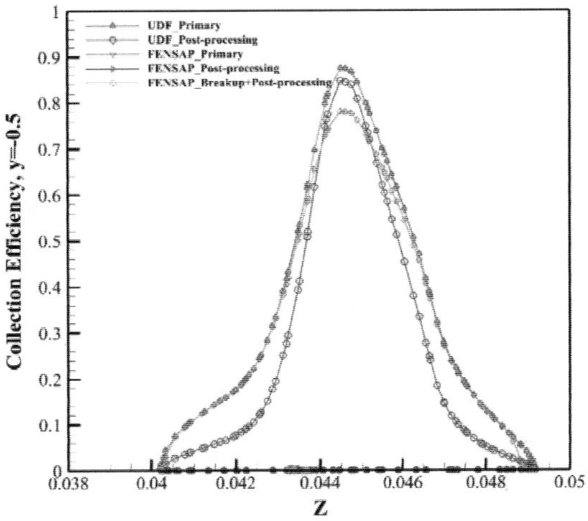

Figure 8. Comparison curve of collection rates at the section.(y=-0.5m)

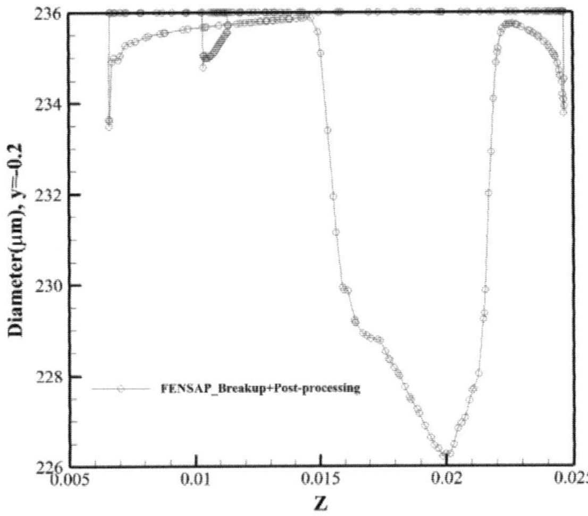

Figure 9. Diameter curve at slicing point (y=-0.2m)

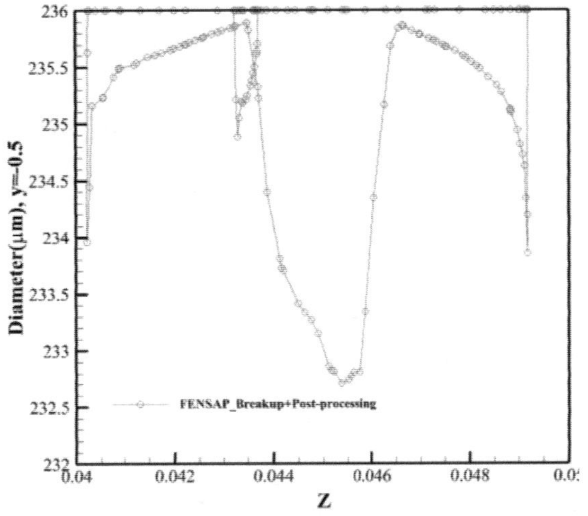

Figure 10. Diameter curve at slicing point (y=-0.5m)

The collection rates were further compared at wing sections y=-0.2m and y=-0.5m, with the section locations shown in Figure 6. Additionally, corresponding test cases were calculated using FENSAP software for comparison, where "FENSAP_Primary" represents results calculated without activating the SLD model, "FENSAP_Post-processing" shows results obtained through post-processing splashing in FENSAP using the "Wright" splashing model, and "FENSAP_Breakup+Post-processing" indicates results calculated by simultaneously activating the breakup model and post-processing. Figure 7 and 8 shows the comparison curves of collection rates at the sections, with both y=-0.2m and y=-0.5m locations exhibiting identical comparative characteristics. The primary collection rate curves calculated by UDF and FENSAP almost completely overlap. The "FENSAP_Post-processing" results appear more conservative compared to "UDF_Post-processing", likely because FENSAP uses a higher splashing threshold where splashing only occurs near the peak collection rate while the curve flanks remain unaffected, resulting in reduced collection rates. Furthermore, near the peak region, "FENSAP_Post-processing" shows greater splashing mass loss than "UDF_Post-processing", which exhibits almost no mass loss at the peak. The curves for "FENSAP_Post-processing" and "FENSAP_Breakup+Post-processing" are nearly identical because droplet breakup has minimal effect on diameter changes, as evidenced by the diameter curves at the sections shown in Figure 9 and 10, where most droplets impacting the wall exceed 226μm, thus having limited influence on splashing post-processing. Therefore, it can be acceptable in some cases to neglect breakup effects when performing large droplet analysis calculations.

The three-dimensional re-entrainment calculation for large droplets is implemented through customized UDF programming. The computational methodology proceeds as follows: Building upon the Eulerian droplet field solution, all wall-adjacent droplets are evaluated against the splashing threshold criterion to determine splash occurrence. At surface elements where splashing is predicted, Lagrangian droplet particles are injected with initial velocities derived from both the local Eulerian droplet velocity field and the splashing model's kinematic relationships. This study employs the following splashing criterion formula:

$$K_{ctr} = \frac{\Lambda^{-0.375}\sqrt{K}}{(\sin\theta_o)^{1.25}} > 200 \tag{8}$$

Where K represents the splash parameter proposed by Mundo to quantify droplet impact energy, defined as:

$$K = \left(Oh^{-0.4}We_n\right)^{0.625} = \left(\frac{\rho_w^3 d_w^3 u_{w,n}^5}{\sigma_w^2 \mu_w}\right)^{0.25} \tag{9}$$

Particle injection positions are specified at the centroid of triangular surface elements (quadrilateral elements being outside this study's scope), necessitating tessellation of all geometric surfaces into triangular facets.To enhance re-entrainment droplet resolution, splashing surface elements may be subdivided. Each triangular element's edges are partitioned into Num_Edge_Divide equal segments, with connecting nodes generating Num_Edge_Divide × Num_Edge_Divide congruent sub-triangles. Correspondingly, Num_Edge_Divide × Num_Edge_Divide droplet particles are injected at each sub-triangle's centroid, as illustrated in Figure 9. The mass information carried by each particle is recorded via the TP_USER_REAL macro, mathematically expressed as:

$$Mass_d = \frac{f_m \beta_{primary} A_{reinjection}}{Num\_Edge\_Divide^2} \tag{10}$$

Here, μ represents the mass ratio between splashed droplets and incident droplets, $\beta_{primary}$ denotes the wall collection efficiency obtained from primary Eulerian computation, and $A_{reinjection}$ indicates the area of the surface element where particles are injected. When the injected splashing particles deposit on

Journal of Physics: Conference Series **3056** (2025) 012007 doi:10.1088/1742-6596/3056/1/012007

downstream surfaces after trajectory calculation, their contribution to the collection efficiency of the impacted surface element is given by:

$$\beta_{reinjection} = Mass_d / A_{imp} \tag{11}$$

Here, A_{imp} denotes the area of the impact surface element. The mass contribution from droplets undergoing multiple splash events is considered negligible in this study, and thus higher-order splashing is not computed. The initialization of re-entrained droplets - including their quantity, spatial distribution, and velocity field - is implemented through the DEFINE_DPM_INJECTION_INIT macro.

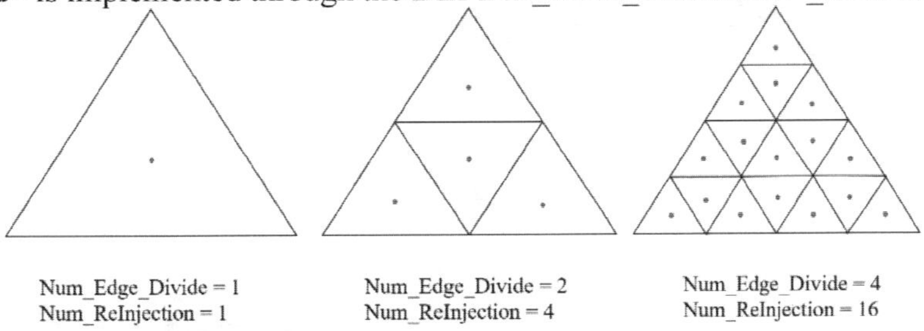

Figure 11. Diagram of water droplet injection into the splash face unit

The wing surface comprises a total of 116,872 face elements, with 55,306 elements reaching the splashing threshold. The fuselage surface contains 31,710 face elements, of which 16,421 meet the splashing criterion. Figure 10 displays the trajectory plot of particles released exclusively from the wing surface (55,306 injected particles), while Figure 13 presents the water collection rate distribution for re-entrant particles on the wing surface. The results demonstrate that the collection rate contribution from wing re-entrant particles is minimal and practically negligible, as the majority are carried away by the airflow with minimal surface deposition. The primary contribution to re-entrant collection efficiency originates from droplet re-entry in the nose region.

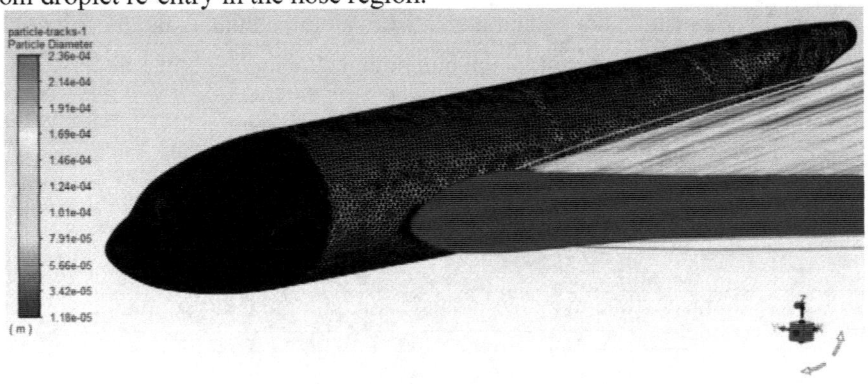

Figure 12. Schematic diagram of the motion trajectory of particles injected on the surface of the wing.

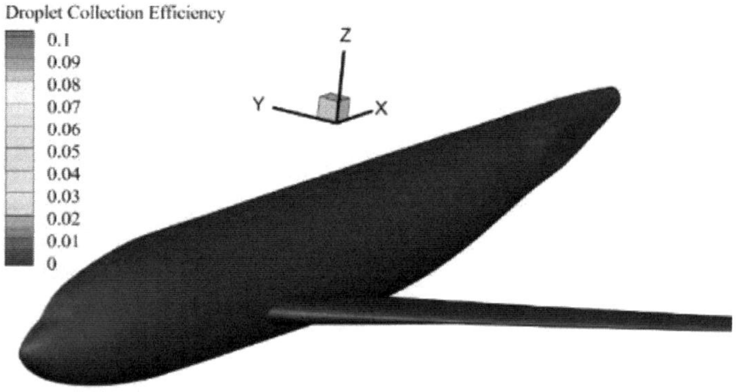

Figure 13. The collection rate of water droplets injected into the surface of the wing.

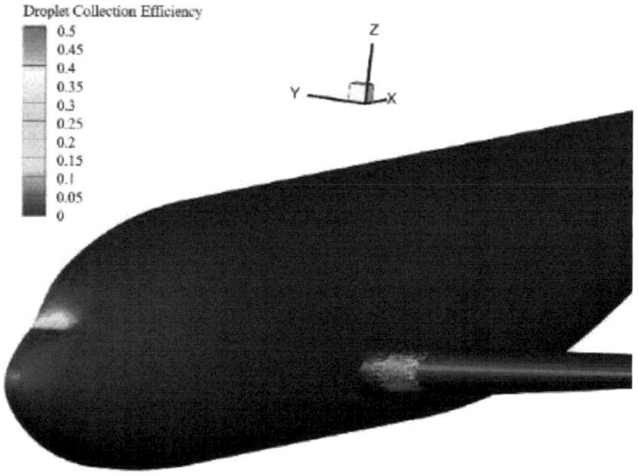

Figure 14. Re-entrance water droplet collection rate, Num_Edge_Divide=1

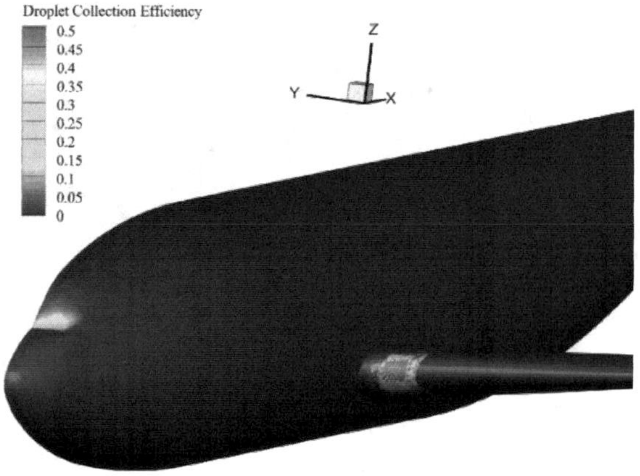

Figure 15. Re-entrance water droplet collection rate, Num_Edge_Divide=5

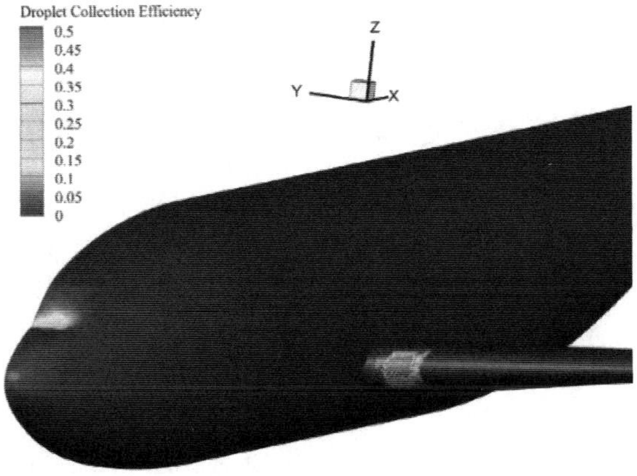

Figure 16. Re-entrance water droplet collection rate, Num_Edge_Divide=10

Figures 14 to 16 respectively present the re-entrant water collection rates on the fuselage surface with Num_Edge_Divide set to 1, 5, and 10, where the numbers of injected and tracked particles are 16,421, 82,105, and 1,642,100 correspondingly. As Num_Edge_Divide increases, the calculation of re-entrant collection efficiency demonstrates improved accuracy.

The results further indicate that significant re-entrant collection occurs primarily in the nose region and wing-root area, with the latter exhibiting particularly severe droplet re-impingement. This phenomenon may lead to local water accumulation and subsequent ice formation. For ice protection systems, targeted mitigation strategies may be required to address this risk.

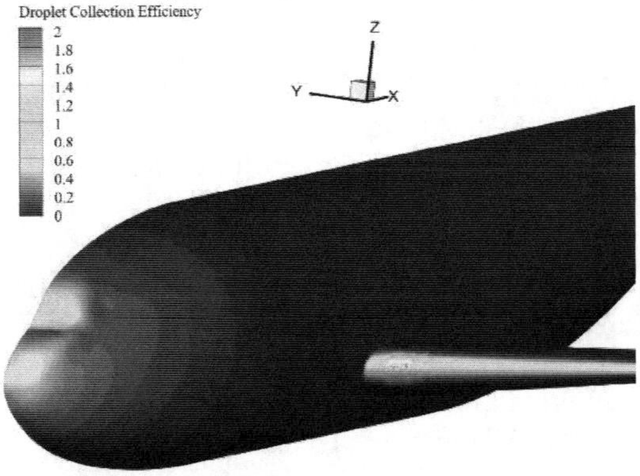

Figure 17. Overall collection rate, Num_Edge_Divide=10

Figure 17 displays the integrated collection efficiency, obtained by superimposing the post-processed collection rates from Figure 5 and the re-entrant collection rates from Figure 16. Due to the influence of re-entrant droplets, certain locations on the nose and wing-root surfaces exhibit collection efficiencies exceeding 1.0.

4. Conclusion

This paper presents the mathematical models for calculating water collection efficiency using both Eulerian and Lagrangian approaches, along with their respective advantages and limitations. Validation of the two-dimensional large droplet Discrete Phase Model demonstrates the code's reliability. Finally, a novel hybrid solution method combining Eulerian and Lagrangian frameworks is proposed.

Comparing the results of FENSAP and UDF calculations, it can be seen that the primary collection rate curves are nearly overlapping, indicating the reliability of the UDF calculations. On the other hand, the results from FENSAP are more conservative, likely due to the higher splash threshold in FENSAP, which leads to a decrease in collection rate. This would lead to an underestimation of actual splashing losses, resulting in overpredicted ice accretion.

The DPM re-entrainment computation results demonstrate that the wing-root region captures additional water droplets due to droplet re-entry, potentially leading to local water accumulation and subsequent ice formation. These findings confirm the capability of the presented DPM code for three-dimensional large droplet (SLD) re-entrainment simulations. Furthermore, the results exhibit improved accuracy in re-entrainment collection efficiency calculations with increasing Num_Edge_Divide refinement.

Acknowledgments

This work was supported by Natural Science Foundation of Shandong Province (ZR2022MA075), and National Natural Science Foundation of China (U1833121).

References

[1] Honsek R and Habashi W G. 2006 FENSAP-ICE: Eulerian Modeling of Droplet Impingement in the SLD Regime of Aircraft Icing
[2] Honsek R 2005 Development of a Three-Dimensional Eulerian Model of Droplet-Wall Interaction Mechanisms
[3] Iuliano E, Mingione G, Petrosino F, et al. 2010 Eulerian modeling of SLD physics towards more realistic aircraft icing simulation
[4] Bilodeau D R, Habashi W G, Fossati M, et al 2015 Eulerian Modeling of Supercooled Large Droplet Splashing and Bouncing. Journal of aircraft 52(5):1611-1624.
[5] Wang C, Chang S, Wu H. Lagrangian 2014 Approach for Simulating Supercooled Large Droplets' Impingement Effect. Journal of Aircraft 52:1-14.
[6] Wang C, Chang S, Leng M, et al 2016 A two-dimensional splashing model for investigating impingement characteristics of supercooled large droplets. International Journal of Multiphase Flow 80:131-149.
[7] Shen X, Tan Y, Rendong Y, et al 2021 Effects of Upstream Component and Air Injection on Water Droplet Impingement Characteristics for Downstream Surfaces. International Journal of Aerospace Engineering 021:1-12.
[8] Han Han, Li Yao, Yin Zifei, et al 2022 Eulerian-Lagrangian Sampling Method of SLD Droplet Size Distribution and its Influence on Ice Shapes[J] Science Technology and Engineering 22(20): 8960-8971.
[9] Bellosta T, Baldan G, Sirianni G, et al. 2023 Lagrangian and Eulerian algorithms for water droplets in in-flight ice accretion[J]. Journal of Computational and Applied Mathematics, , 429:115230.
[10] https://aiaa-dpw.larc.nasa.gov/Workshop2/workshop2.html[Z].

Journal of Physics: Conference Series **3056** (2025) 012008 doi:10.1088/1742-6596/3056/1/012008

Analysis and Verification of Fluid Cooling System Leakage Fault of Manned Spacecraft

Yang FU, Yaoyao ZHU, Zhangpeng MING*, Chunyang NIU, Shigui ZHENG, Dongpu WANG, Xianlin WU

Beijing Institute of Spacecraft System Engineering, Beijing 100094, China

Corresponding author's e-mail: phdming@yeah.net

Abstract. The fluid loop thermal radiator, serving as the primary thermal dissipation equipment for large spacecraft, is installed in the outermost structural layer of the spacecraft. Micro-Meteoroid and orbital debris(MMOD) penetrated the radiator tubing, causing coolant leakage and leading to failure of the fluid cooling system. Taking the Shenzhou manned spacecraft as the research subject, this study establishes an MMOD impact prediction model for the radiator and evaluates its protective capability. The manned spacecraft executed a round-trip mission to the Space Station. The fault modes of the radiator were analyzed, and the full-spacecraft thermal balance tests were carried out. The test results indicate that when the cooling system leakage occurs during the spacecraft's parking with the Space Station, both in the high and low temperature conditions, the equipment and the ambient temperature can remain within acceptable limits. For leakage scenarios during autonomous flight, implementing measures such as reducing spacecraft power dissipation ensures that equipment temperatures stay within operational requirements.

1. Introduction

Between December 2022 and October 2023, three coolant leakage events occurred in the Russian segment of the International Space Station(ISS), specifically:

(1) Coolant leakage from the Soyuz MS-22 manned spacecraft on December 15, 2022[1].

(2) Coolant leakage from the Progress MS-21 cargo spacecraft on February 11, 2023[2].

(3) Coolant leakage from the external backup radiator of the Russian Nauka multipurpose laboratory module on October 9, 2023[3].

All three events involved radiator leaks, with the first two events revealing similar potential leak positions. The third event occurred after the installation of a radiator spare part that had been stored for 13 years.

During the International Space Station (ISS) mission, damage to the station's surface caused by space debris has been widely observed, including impacts on radiators[4]. The fluid circuit design of

Journal of Physics: Conference Series **3056** (2025) 012008 doi:10.1088/1742-6596/3056/1/012008

the thermal control subsystem in the Shenzhou series of manned spacecraft shares similarities with the Soyuz and Progress spacecraft, as both dissipate heat into space with radiators[4]. When the exposed radiator pipelines are pierced by debris, the coolant leakage can occur. This indicates that the Shenzhou series of manned spacecraft shares the same leak failure mode as the Russian Soyuz and Progress spacecraft [5].

To address the threat of space debris to the Shenzhou spacecraft, several analyses and studies have been conducted. In the 1990s, NASA's Marshall Space Flight Center developed the Manned Spacecraft Crew Survivability Computer Model, which was the first to assess the non-catastrophic failure probability of manned spacecraft and astronauts in MMOD (Micro-Meteoroid and Orbital Debris) environments[6]. In China, Qu Guangji and others were among the early researchers to conduct modeling and numerical simulations of space debris impacts [7]. Yu Qingbo and colleagues analyzed the influence of orbital parameters on the risk of space debris collisions with manned spacecraft, focusing on the impact on the structural hull [8]. Han Zengyao and others have developed practical software for assessing debris impact risks[9]. Huang Lei and others have conducted in-depth analyses of radiator leaks caused by debris impacts, modeling the relationship between leak size and fluid leakage rate[10]. Their research findings indicate that when the leak diameter exceeds 1 mm, the pressure in the fluid drops rapidly, potentially leading to a loss of cooling capability within minutes. Thus, radiator leaks pose a significant risk to the thermal control functionality of the spacecraft.

This paper analyzes the failure conditions of the fluid cooling system in the Shenzhou series of manned spacecraft and validates these conditions through full-spacecraft thermal balance tests.

2. Introduction to the Shenzhou Spacecraft fluid cooling system

The fluid circuit system is the core of the spacecraft's thermal control system, integrating heat collection, transmission, dissipation, and distribution. It is an active thermal control system with robust thermal and humidity regulation capabilities, playing a critical role in precisely controlling the temperature and humidity of the spacecraft under various multi-task and multi-mode conditions.

The fluid circuit of the Shenzhou manned spacecraft can be divided into two loops based on their functions: the inner loop and the outer loop. The inner loop primarily located in the orbital and re-entry modules, is responsible for cooling and dehumidifying the cabin air and cooling equipment. The heat collected by the inner loop is transferred to the outer loop via a heat exchanger. The outer loop, located in the propulsion module, is responsible for cooling non-hermetically sealed equipment, collecting heat from the inner loop, and dissipating heat to space. It also provides supplementary heating for certain areas of the spacecraft. The schematic diagram of the fluid circuit is shown below.

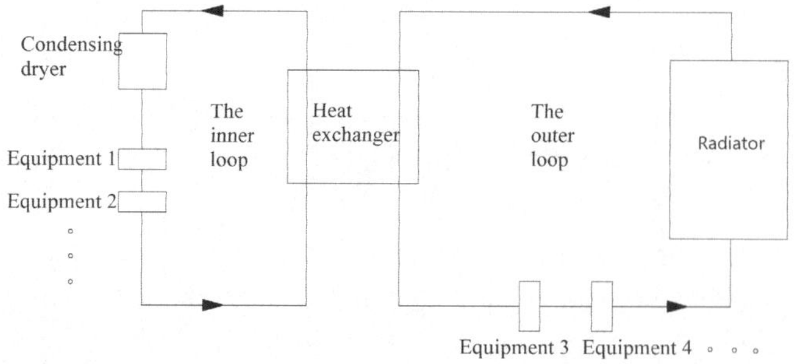

Figure 1. The schematic diagram of the fluid circuit in Shenzhou manned spacecraft

3. Risk Assessment of Radiator Impact and Leakage for the Shenzhou Spacecraft

The radiators of the Shenzhou manned spacecraft are installed on the exterior of the propulsion module, mainly functioning to dissipate the waste heat of spacecraft into space. The radiator pipelines are distributed across the radiator surface.

When the radiator pipelines are pierced by space debris, coolant leakage can occur. This section assesses the risk of radiator damage from space debris during the spacecraft missions.

The orbit of the manned spaceship is a circular orbit with an average orbital altitude of approximately 390 km. The space debris environment is described by the internationally recognized ORDEM2000 [11], with an average density of 2.8 g/cm^3. The micrometeoroid model adopts the internationally recognized NASA SSP 30425 [12], with an average density of 1.0 g/cm^3.

The Shenzhou spacecraft primarily experiences three mission phases: autonomous flight and docking, parking, and return. During autonomous flight and docking, the spacecraft adopts a three-axis earth-pointing attitude. During the docking and parking phase, the long-term attitude of spacecraft is forward flight, with the spacecraft docked at the radial or forward node of the space station. The probability of space debris impact on the radiator is simulated as follows, and the results are presented in Table 1 and Table 2.

Table 1: Assessment of the impact probability on radiators of manned spacecraft(one year)

Parking position	Debris diameter (mm)	Impact number	Impact probability
Forward parking	>0.1	1.0875E+00	6.6293E-01
	>0.5	1.8829E-02	1.8653E-02
	>1.0	1.6680E-03	1.6666E-03
Radial parking	>0.1	1.2245E+00	7.0609E-01
	>0.5	2.1033E-02	2.0813E-02
	>1.0	1.8644E-03	1.8627E-03

Table 2. Assessment of the penetration probability on radiators of manned spacecraft (one year)

Docking position	Penetration number	Penetration probability
Forward docking	5.3456E-02	5.2052E-02
Radial docking	7.8627E-02	7.5615E-02

According to the results of the calculation, the penetration probability on radiators of Shenzhou manned spacecraft is still high. During the radial parking phase, the probability of radiator impact is higher than that of forward parking or autonomous flight (results of autonomous flight can be approximated to forward parking assessments). To ensure the safety of the spacecraft and astronauts, appropriate measures must be taken.

4. Improvements for Leakage Failures

Multiple design improvements were made to the Shenzhou manned spacecraft to address leak failures. There are two main aspects.

(1) For the equipment in the non-sealed cabin, the fluid cooling system not only provides cooling but also heating at low temperatures. Therefore, the high and low temperature of the equipment

Journal of Physics: Conference Series **3056** (2025) 012008 doi:10.1088/1742-6596/3056/1/012008

should be considered simultaneously in case of complete failure of the fluid cooling system. To mitigate low-temperature risks, electric heating measures have been added. To mitigate high-temperature risks, passive heat dissipation measures and power reduction measures have been implemented.

(2) For the equipment in the sealed cabin, since there is air in the cabin, equipment temperatures will not drop excessively. Therefore, the primary concern is high-temperature issues when the fluid cooling system fails. For high-temperature equipment, fan have been added to carry out forced convection heat dissipation, combined with power reduction measures.

5. Leakage Failure Analysis

5.1. Analysis of Parking Phases

During parking phases, the Shenzhou spacecraft can remain in orbit for six months or longer, during which the risk of penetration of the radiator pipeline significantly increases. If a radiator pipeline is pierced, the coolant is expected to leak completely, leading to a total loss of the functionality of the fluid cooling system.

For air temperature, the space station provides thermal support to the sealed cabin of the Shenzhou spacecraft through air circulation. Thus, the air temperature and humidity in the sealed cabin are primarily maintained by air supply from the space station.

For equipment temperature, the sealed cabin benefits from space station air circulation, ensuring the environmental temperature and the equipment temperature. And the high-power equipment is in a shutdown state, so the risk of high temperature for the equipment is low. Non-sealed cabin equipment relying on the fluid cooling system heating (e.g., lithium battery packs) may face low-temperature risks.

Based on the above analysis, the main risk during the parking phase is the low-temperature risk to the equipment.

5.2. Analysis of Autonomous Flight Phases

The autonomous flight of the Shenzhou manned spacecraft is mainly divided into two phases: the autonomous flight and docking phase, and the return phase.

During autonomous flight and docking phases, the Shenzhou spacecraft operates for a relatively short duration, and adopting a rapid docking mode requires only a few hours to complete. Thus, the probability of space debris impact during this phase is relatively low. And the leak hole is small, the coolant will leak over a period of time. During this interval, the compensator in the fluid cooling system will compensate for coolant loss, and the fluid cooling system will retain partial cooling capability. As the leak worsens, the fluid cooling system will gradually lose its cooling capacity.

If a leakage failure occurs during the parking phase, the fluid cooling system will have completely lost its cooling capability during the return phase.

During the autonomous flight phase, high-power consumption equipment of the spacecraft are activated, and astronauts in the sealed cabin generate body heat. Therefore, both the equipment and the air have high-temperature risks.

6. Analysis and Validation of Fault Condition

Based on the above analysis, both high- and low-temperature fault conditions should be assessed during the parking phase; while dynamic thermal conditions must be evaluated during the autonomous flight phase. The subsequent analysis focuses on equipment temperature and air temperature, incorporating data from full-spacecraft thermal balance tests.

6.1. High-Temperature and Low-Temperature Fault Conditions During Parking

The high-temperature parking condition was selected with a solar angle of $|\beta| = 65°$, with the spacecraft docked at the forward node of the space station. In this configuration, the maximum external heat flux occurs in quadrant II or IV of the manned spacecraft. The low-temperature parking condition was selected with a solar angle of $\beta = 0°$, with the spacecraft docked at the radial node of the space station, resulting in the most severe shielding from the space station and the minimum external heat flux.

Full-spacecraft thermal balance test results of air temperature are shown in Table 3. In the high-temperature fault condition, the air temperature of the return cabin is slightly higher, but all temperatures meet the safety requirements for the astronauts.

Table 3: Air temperature in the sealed cabin under high and low temperature fault conditions (Temperature values represent the average of three measurement points, with the maximum temperature recorded during the dynamic simulation process.)

	Parking under high-temperature fault condition ($|\beta| = 65°$)	Parking under low-temperature fault condition ($\beta = 0°$)
Return cabin air temperature(℃)	26.5	24.1
Orbital cabin air temperature(℃)	24.9	21.6

Equipment temperatures are shown in Table 4. The temperature of all equipment are good. Under a low-temperature fault condition, when the electric heating is activated, the temperature requirements of the lithium batteries can be met.

Table 4. High-Power Equipment Temperatures Under High and Low Temperature Fault Conditions

Equipment name	Normal operating temperature range (℃)	High-temperature condition (℃)		Low-temperature condition (℃)	
		Normal condition	Fault condition	Normal condition	Fault condition
Gyro equipment	-5~50	19.0	28.0	18.3	22.7
Digital equipment	-10~50	18.1	30.1	17.4	24.0
GNC equipment	-15~50	20.4	31.0	19.9	25.5
Lithium battery pack a	10~35	17.5	22.3	13.0	11.3*
Lithium battery pack b	10~35	16.7	20.6	12.9	11.3*
Lithium battery pack c	10~35	16.2	19.2	12.6	11.3*
Antenna equipment a	-5~50	18.2	37.4	17.6	31.7
Antenna equipment b	-5~40	17.6	26.8	16.7	20.6

*When the temperature drops below 11.3 ° C, heating system is activated, and the equipment temperature is maintained via electric heating.

Journal of Physics: Conference Series **3056** (2025) 012008 doi:10.1088/1742-6596/3056/1/012008

The above verification results demonstrate that both ambient air and equipment temperatures can be effectively maintained, thereby ensuring the operational safety of the Shenzhou spacecraft during the parking phase.

6.2. *Autonomous Flight Fault Conditions*

For faults during the autonomous flight and docking phase, the corresponding normal steady-state temperature is chosen as the starting point, with a 1:1 dynamic simulation conducted to evaluate the worst-case scenario. To minimize equipment temperatures, a full-spacecraft power reduction mode is adopted. Since the test conditions start from the corresponding normal operating conditions, which differ slightly from the actual operating conditions, primarily because many equipment items are activated during testing, whereas some remain inactive in the real pre-launch phase. Consequently, the test conditions are more severe compared to the in-orbit conditions.

The full-spacecraft thermal balance tests included two simulations of autonomous flight and docking conditions. One simulation was for high-temperature fault conditions with a solar angle of 40°, and the other was for low-temperature fault conditions with a solar angle of 0°. According to the results in Tables 5 and 6, the equipment temperatures met the technical requirements, but the air temperature was higher than expected. When the solar angle was 0° compared to 40°, the air temperature was approximately 3°C lower. External heat flux had a significant impact on the overall temperature of the spacecraft.

For the return phase faults, the high-temperature docking fault condition was chosen as the starting point, but some measures were taken to reduce the initial temperature of the entire spacecraft. The start point was the separation of the manned spacecraft, and the test was conducted with a 1:1 dynamic simulation. The test was conducted with a 1:1 dynamic flight simulation assessment, which is closer to the actual in-orbit conditions. To further reduce the equipment temperature, a power-reducing mode for the entire spacecraft was also adopted. According to the results in Tables 5 and 6, the equipment temperatures met the technical requirements, but the air temperature was higher than expected.

Table 5. High-power equipment temperatures under autonomous flight fault conditions (℃)

Equipment name	Normal operating temperature range	Autonomous flight and docking high-temperature fault conditions ($\beta = 40°$)	Autonomous flight and docking low-temperature fault conditions ($\beta = 0°$)	Return high-temperature fault conditions ($\beta = 40°$)
Gyro equipment	-5~50	38.2	37.0	30.4
Digital equipment	-10~50	32.6	30.2	29.6
GNC equipment	-15~50	34.1	32.2	31.0
Lithium battery pack a	10~35	23.7	21.7	22.5
Lithium battery pack b	10~35	20.8	19.2	18.5
Lithium battery pack c	10~35	19.4	18.1	17.0
Antenna equipment a	-5~50	39.8	36.4	33.4

Journal of Physics: Conference Series **3056** (2025) 012008 doi:10.1088/1742-6596/3056/1/012008

Antenna equipment b	-5~40	31.3	29.4	30.1

Table 6. Air temperature in the sealed cabin under autonomous flight fault conditions（Temperature values represent the average of three measurement points, with the maximum temperature recorded during the dynamic simulation process.）

	Return cabin highest air (℃)	Orbital cabin highest air (℃)
Autonomous flight and docking high-temperature fault condition (β = 40°)	34.3	26.5
Autonomous flight and docking low-temperature fault condition (β = 0°)	30.9	25.2
Return high-temperature fault condition (β= 40°)	32.9	20.3

6.3. *Fault Countermeasures*

For autonomous flight and docking phases, equipment temperatures under fault conditions meet the requirements of flight tasks. However, air temperatures are elevated, posing a high-temperature risk to personnel. Given the short duration of autonomous flight and docking (only a few hours), the probability of faults is relatively low. Additionally, after a leakage fault occurs, there may still be some cooling capacity for a short period. Therefore, subsequent flight tasks can be carried out based on the timing of the fault and the extent of the leakage.

During re-entry, the air temperature of the return module is expected to rise by an average of 4.3°C due to aerodynamic heating. Equipment temperatures will also increase by the same extent. Based on the analysis, equipment temperatures will still meet mission requirements. The air temperature of the return module is expected to reach 37.2°C. To mitigate this, the spacecraft can return under conditions with lower external heat flux, such as a solar angle of 0°. This would reduce the air temperature of the return module to 34.2°C, which is still high and exceeds the comfort temperature range for astronauts.

Additionally, the air temperature in the orbital module is very low, only 20°C. Due to the closure of the re-entry module's hatch, the heat cannot be balanced between the two modules. Subsequently, it can be considered to conduct full-spacecraft thermal balance tests to facilitate air exchange between the two modules as much as possible, in order to lower the air temperature in the re-entry module. It is expected that this will bring the air temperature in the re-entry module below 30°C.

7. Conclusion

This paper conducted an in-depth analysis of the failure scenarios of the fluid cooling system in the Shenzhou series of manned spacecraft and validated these conditions through full-spacecraft thermal balance tests.

（1） The assessment of radiator impact and penetration probabilities identified a significant risk of radiator pipeline penetration in a manned spacecraft.

（2） To address leakage failures, improvements were proposed for both non-sealed cabin and sealed cabin equipment, including electric heating, forced convection cooling, and power reduction measures. These improvements effectively mitigated high- and low-temperature risks.

Journal of Physics: Conference Series **3056** (2025) 012008 doi:10.1088/1742-6596/3056/1/012008

(3) Full-spacecraft thermal balance tests confirmed that during parking fault conditions, both equipment and air temperatures met technical requirements, ensuring safe parking. During autonomous flight fault conditions, equipment temperatures were within acceptable limits, but air temperatures were elevated. Solutions were proposed to address high air temperatures.

This research provides critical references for risk assessment and flight control strategies related to fluid cooling system failures in the Shenzhou manned spacecraft. It also offers insights for optimizing thermal control system design, enhancing space debris protection capabilities, and improving thermal management strategies.

Reference

[1] Spacewalk Cancelled, Mission Controllers Evaluate Leak on Soyuz. http://www.nasa.gov/blogs/spacestation/2022/12/14/spacewalk-cancelled-mission-controllers-evaluate-leak-on-soyuz.html

[2] Progress Resupply Undocks form station, Ends Cargo Mission. http://www.nasa.gov/blogs/spacestation/2023/02/17/ progress-resupply-undocks-form-station-ends-cargo-mission.html

[3] International space station astronauts nauka ammonia leak no danger. https://www.space.com/international-space-station-astronauts-nauka-ammonia-leak-no-danger

[4] James L. Hyde, Eric L. Christiansen, Dana M. Lear 2019 Observations of MMOD Impact Damage to the ISS. International Orbital Debris Conference(Sugarland)p 9-12

[5] HUANG J, FAN Y, YU S, et al. 2009 On-orbit Performance Evaluation of Single-phase Fluid Loop System for Shenzhou-7 Spaceship Spacecraft Engineering 18(4) p 37-43

[6] Joel W. 1994 Quantitative analysis of manned spacecraft crew survivability from orbital debris penetration Space Programs and Technologies Conference(Huntsville) p 27-29

[7] YU Q, WANG H, WANG H, et al. 2010 Effect of Orbit Parameters on Manned Spacecraft Impact Risk Against Orbital Debris Science & Technology Review 28 (2) p 34-38

[8] QU G, HAN engyao. 2022 Dynamical Modeling and Numerical Simulation of Hyperveloctiy Space Debris Impact Chinses Space Science and Technology 22(5) p 26-30.

[9] HUANG L, YANG M, CHE B, et al. 2024 Analysis and Verification of Puncture Leakage Characteristics of Spacecraft Heat Radiator by MMOD Spacecraft Engineering 33(2) p 139-143

[10] HAN Z, ZHENG S, YAN J, et al. 2005 Development, Calibration and Application of Space Debris Impact Probability Analysis Software Journal of Astronautics 26 (2) p 228-231

[11] J-C Lion, et al, 2002 The New NASA Orbital Debris Enineering Model ORDEM2000, NASA TP -2002 - 210780

[12] NASA, 1993 Meteoroid Environment Model, Space Sation Program Natural Enviroment Definition for Design,SSP-3025, Rev.B

Prediction of Frost Formation on Aircraft Wing Skin Before Takeoff in Winter

Miao Zheng[1], Weijian Chen[1*[0000-0001-6776-4492]], Xiao Han[2], Ning Guan[3]

[1] Key Laboratory of Aircraft Environment Control and Life Support, MIIT, Nanjing University of Aeronautics & Astronautics, Nanjing, China, 210016

[2] Avic Aerospace life-support Industries,Ltd, Xiangyang, Hubei Province, China, 441003

[3] College of Aeronautics, Shandong Jiaotong University, No.5001, Haitang Road, Changqing University Science and Technology Park, Jinan, China, 250037

*Corresponding author's e-mail: weijian_chen@nuaa.edu.cn

Abstract. Numerical simulation of frost formation on aircraft wing skins with internal fuel tanks was conducted based on a frost growth prediction model, focusing on the extreme temperature rise phase during winter pre-takeoff conditions. The simulation employed FLUENT software with UDF programming under selected operating conditions to obtain temperature distributions on both upper and lower wing surfaces, frost accumulation amounts, condensed water content, and contour maps of frost layer thickness and density distribution. Analysis of the simulation results indicates that during winter pre-takeoff conditions, wings in fully-fueled state exhibit greater frost accumulation and thicker frost layers, while wings in empty-fuel state demonstrate higher frost density. These findings provide valuable insights for aircraft anti-icing strategies in cold weather operations.

1. Introduction

Formation of frost represents a widespread natural physical phenomenon. However, when it occurs in industrial production, it becomes a significant hazard that requires urgent resolution, especially in the aerospace field[1], where the working environment is complex .On 21 November 2004, a CRJ-200 aircraft crashed near Baotou Airport, resulting in 55 fatalities. The accident investigation team concluded that frost-conducive weather conditions existed during the aircraft's overnight stay at Baotou Airport, and frost had likely formed on the wings. Notably, no de-icing procedures were conducted prior to takeoff.

Researchers have studied frost formation for decades.Hayashi[2] observed the complete frost formation process using a microscopic camera and divided cold surface frosting into three stages: crystal growth, layer growth, and mature stage. O'neal[3]investigated the effects of Reynolds number, air humidity, air temperature, and plate temperature on frost growth and density on vertical plates under parallel flow conditions.Webb[4] proposed a new numerical model for frost deposition and growth.Lee[5] developed governing equations for both frost layer and humid air, and proposed a frost growth prediction model under turbulent flow conditions.

Therefore, it is essential to know the hazards of wing frost formation before aircraft takeoff [6]and use frost prediction models to estimate key parameters such as frost accumulation[7] and frost density[8]

Journal of Physics: Conference Series **3056** (2025) 012009 doi:10.1088/1742-6596/3056/1/012009

on the wing surface.

Nomenclature

Symbol			
D	Diffusion coefficient ($m^2 \cdot s^{-1}$)	P	Pressure (Pa)
α_f	Absorption coefficient(s^{-1})	q	Heat flux ($W \cdot m^{-2}$)
h_d	Mass transfer coefficient ($m \cdot s^{-1}$)	λ	Thermal conductivity($W \cdot m^{-1} \cdot K^{-1}$)
h_h	Heat transfer coefficient ($W \cdot m^{-2} \cdot K^{-1}$)	T	Temperature (K)
ρ	Density ($kg \cdot m^{-3}$)	v	Velocity ($m \cdot s^{-1}$)
L_h	Sublimation latent heat ($J \cdot kg^{-1}$)	w	Humidity ratio ($kg \cdot kg_a^{-1}$)
m	Mass flux ($kg \cdot m^{-2} \cdot s^{-1}$)	z	Thickness (m)
Subscript			
a	Moist air	p	Cold surface
da	Dry air	sen	Sensible heat
f	Frost	t	Time
f,s	Frost surface	tot	Total
f_0	Initial frost layer	w	Water vapor
lat	Latent heat	w,sat	Saturated water vapor

2. Frost Formation Mathematical Model on Cold Surfaces

During the computational process, key parameters including air temperature, water vapor mass fraction, and cold surface temperature are extracted at each time iteration to accurately simulate the coupled heat and mass transfer process of frost formation.

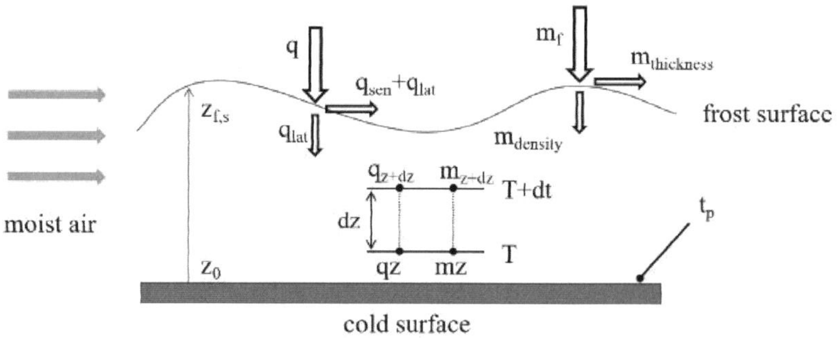

Figure 1. Mathematical model diagram of frost growth

2.1. *Heat and Mass Transfer Processes at Frost Layer Surface*

When humid air passes over a subcooled surface, the net mass transfer rate of water vapor from the air stream to the accumulating frost may be formulated as:

$$m_f = h_d \rho_a (w_a - w_{f,s}) \tag{1}$$

This transfer process consists of dual mechanisms, where a fraction of water vapor experiences instantaneous deposition on the frost interface, contributing to layer thickness growth, while another portion diffuses into the frost interior where it condenses, thereby increasing frost density. The coupled transport process can be quantitatively described by the following expression:

$$m_f = m_{f,thickness} + m_{f,density} = \frac{d}{dt}(z_f \rho_f) = \rho_f \frac{dz_f}{dt} + z_f \frac{d\rho_f}{dt} \tag{2}$$

Two mechanisms govern the heat transfer process between humid air and frost deposits: convective sensible heat exchange constitutes the primary component ,while the other part corresponds to the latent heat released during water vapor deposition onto the frost layer. This thermal exchange process can be formulated as:

$$q_{tot} = q_{sen} + q_{lat} = h_h (T_a - T_{f,s}) + h_d \rho_a (w_a - w_{f,s}) L_h \tag{3}$$

2.2. *Diffusion and Solidification Processes of Water Vapor within Frost Layers*

The diffusion equation governing water vapor transport within the frost layer can be expressed as:

$$D \frac{d^2 \rho_w}{dz^2} = \alpha_f \rho_w \tag{4}$$

Boundary conditionss include: saturation vapor density determined by the plate's local temperature at the cooling surface, coupled with a vanishing vapor density gradient in the normal direction:

$$\left.\begin{array}{l} \rho_w = \rho_{w,sat}(T_p) \\ \dfrac{d\rho_w}{dz} = 0 \end{array}\right\} z = 0 \tag{5}$$

The frost boundary maintains saturated vapor density relative to its surface temperature. Solving the diffusion equation with these boundary specifications yields the frost density distribution along the vertical axis, from which the contributing vapor flux for frost accumulation can be computed:

$$m_{f,density} = D \left.\frac{d\rho_w}{d_z}\right|_{f,s} = \int_{z=0}^{z=z_f} \alpha_f \rho_w dz \tag{6}$$

The diffusion of moist air into the frost layer interior is modeled by accounting for phase change effects through treatment of the latent heat as an internal heat source. Under the quasi-steady-state assumption for frost formation, the time-dependent term may be omitted:

$$\lambda_f \frac{d^2 T}{dz^2} = -\alpha_f L_h \rho_w \tag{7}$$

Under the condition of identical initial temperatures at the frost-substrate interface, the following equations govern the determination of absorption characteristics and thermal profiles:

$$\alpha_f = D \cdot \left[\frac{1}{z_f} \cosh^{-1} \left\{ \frac{\rho_{w,sat}(T_{f,s})}{\rho_{w,sat}(T_p)} \right\} \right]^2 \tag{8}$$

$$T_{f,s} = T_p + \frac{q_{tot}}{\lambda_f} z_f - \frac{D L_h \rho_{w,sat}(T_p)}{\lambda_f} (\cosh(\sqrt{\frac{\alpha_f}{D}} z_f) - 1) \tag{9}$$

2.3. *Temporal Evolution of Frost Layer Thickness and Density*

During frost formation, the frost mass accumulated during each computational time step can be determined by Equation (2). The mass contribution to density increase is quantified by Equation (6). Therefore, the residual mass responsible for thickness growth can be derived as specified below:

$$m_{f,thickness} = m_f - m_{f,density} \tag{10}$$

Accordingly, the temporal variations in frost properties (density and thickness) are computable from the mass deposition rate. This dynamic evolution is mathematically described by the following coupled system:

$$z_f \big|_{t+\Delta t} = z_f \big|_t + \left.\frac{m_{f,thickness}}{\rho_f}\right|_t \Delta t \tag{11}$$

$$\rho_f\Big|_{t+\Delta t} = \rho_f\Big|_t + \frac{m_{f,density}}{z_f}\Bigg|_t \Delta t \tag{12}$$

3. Model and Boundary Conditions

During overnight parking at northern airports in winter, aircraft wings and internal fuel tanks gradually cool through heat dissipation when ambient temperatures drop below -40°C. As temperatures rise toward dawn before takeoff, increasing environmental humidity creates frost formation conditions. When the air's moisture content exceeds the saturation humidity corresponding to the wing surface temperature, frost will develop on wing surfaces.

3.1. Wing Mesh and Material Configuration

This section simulates dynamic frost formation on wing surfaces under coupled multi-physics environmental conditions. The wing is positioned at the center of the flow field, with Figure 2 illustrating the computational mesh of the external flow domain around the wing. Through mesh independence verification, the total number of grid elements has been determined as 1.442 million.

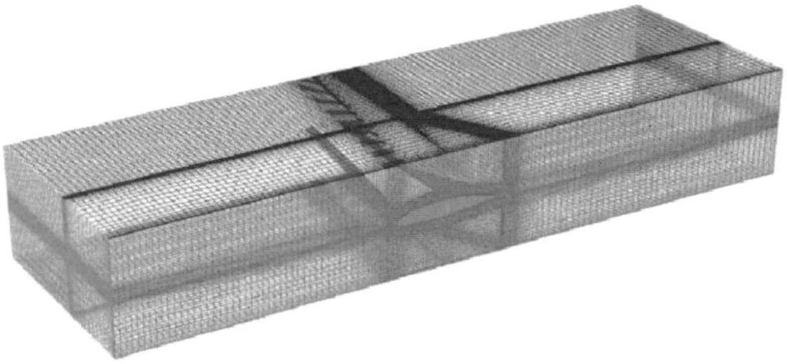

Figure 2. Wing External Flow Field Mesh Schematic

Figure 3 shows the mesh model schematic of the wing with internal fuel tanks, where the yellow areas indicate the tank locations. To reduce computational costs, the modeling process omitted the wing's complex internal truss structures. All internal cavities outside the fuel tank regions were replaced with incompressible gas domains.

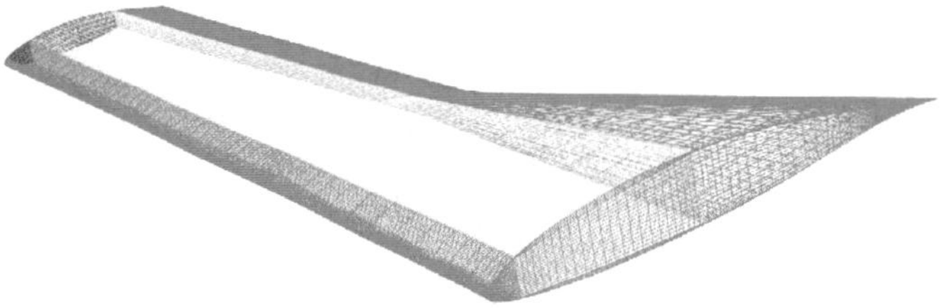

Figure 3. Mesh Schematic of Wing and Fuel Tank Model

This study investigates frost accumulation on wing surfaces under natural convection during ground operations. The numerical approach models the internal wing spaces as a domain of incompressible ideal gas.

3.2. Environmental parameter settings and boundary condition

The simulations model ground operations over 3600 seconds. Ambient pressure is set to 101,325 Pa National climate data shows northern China's winter nighttime temperatures can drop to -40°C, warming to around -20°C by morning. The ambient temperature rises from -40°C to -20°C during the 3600-second period.

Journal of Physics: Conference Series **3056** (2025) 012009 doi:10.1088/1742-6596/3056/1/012009

The wall boundary conditions are specified in Table 1:

Table 1. Boundary condition setup for aircraft wing frost simulation

Zone	Boundary conditions	Settings
inlet	Velocity inlet	$v = 0.1$ m/s
outlet	Pressure outlet	pressure = 101325 Pa
Upper and lower walls	Adiabatic wall	heat flux = 0 w/m^2
Side walls	Symmetry wall	Symmetry
Wing surface	Coupled boundary condition	Thickness = 2mm
Tank surface	Coupled boundary condition	Thickness = 15mm
Oil-air interface	Coupled boundary condition	Thickness = 0mm

The wing frost simulation in this study assumes ground conditions with natural convection flow. The inlet velocity is set to 0.1 m/s. The outlet pressure equals atmospheric pressure. The wing wall and fuel tank wall are coupled with the external fluid.

4. Result

The simulation analyzes wing frost formation before takeoff under two fuel conditions: wing skin frost with full fuel tanks and wing skin frost with empty fuel tanks.

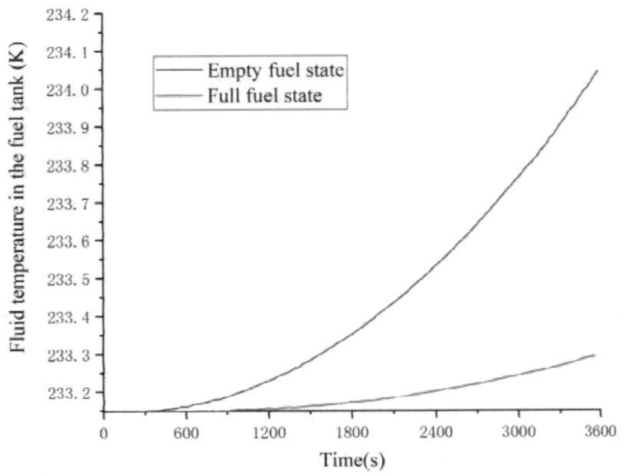

Figure 4. Fuel temperature diagrams under two conditions

In the fully fueled condition, the internal domain of the wing fuel tank is set as kerosene-liquid, in the empty condition, it is defined as air. Figure 4 shows the average temperature variation of the internal domain over time under both conditions. Notably, the fluid temperature in the empty tank is higher than that in the fully fueled tank. This phenomenon occurs because the specific heat capacity of aviation kerosene is significantly greater than that of air.

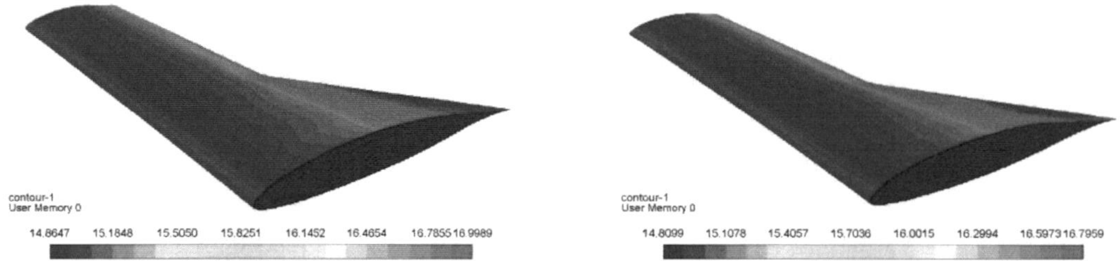

(a) Full fuel state (b) Empty fuel state

Figure 5. Frost accumulation on wing surfaces at 3600s under both conditions

Figure 5 shows the 3600-second frost distribution patterns on wing surfaces for both full and empty fuel tank conditions. The frost layer distribution remains consistent between scenarios, with peak accumulation occurring near the fuel tank on upper and lower wing surfaces. Frost thickness progressively decreases toward the leading and trailing edges, reaching minimum levels at the trailing edge.

The wing surface near the fuel tank maintains the lowest temperatures. When ambient temperature rises, these areas first reach the humid air's dew point, starting frost formation. The trailing edge remains warmer due to its distance from the fuel tank's thermal effects. This temperature difference results in later frost formation, shorter growth time, and less frost accumulation at the trailing edge compared to other wing areas.

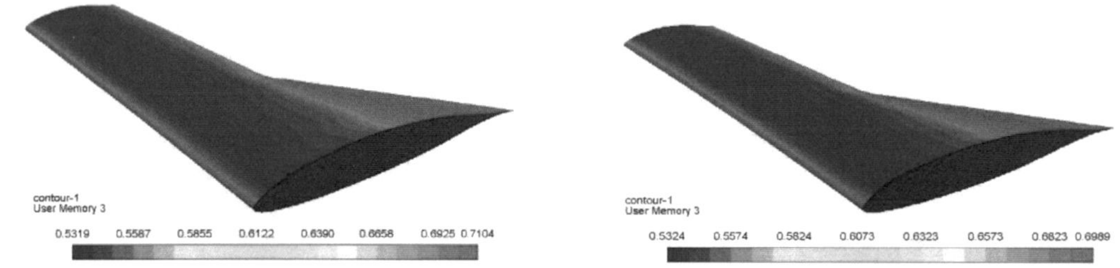

(a) Full fuel state (b) Empty fuel state

Figure 6. Frost thickness on wing surfaces at 3600s under both conditions

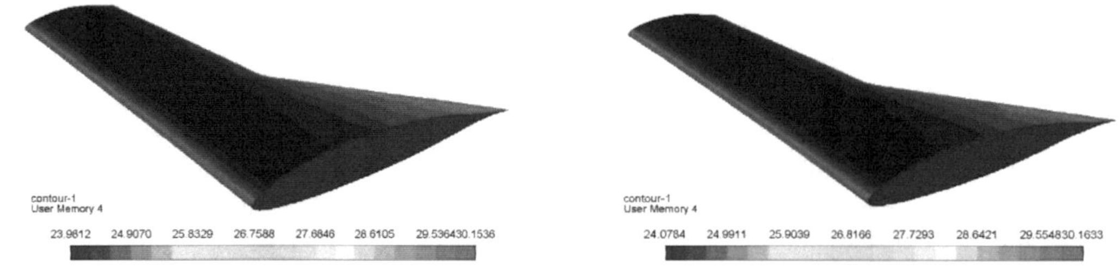

(a) Full fuel state (b) Empty fuel state

Figure 7. Frost density on wing surfaces at 3600s under both conditions

Figure 6 shows the frost thickness distribution at 3600s for both conditions. The thickest frost forms above the fuel tank, with thinner layers at the leading and trailing edges. The leading-edge frost interacts directly with incoming airflow, promoting heat exchange with moist air. This raises surface temperatures, driving water vapor into the frost layer through partial pressure effects. The vapor then changes phase inside the frost, increasing its density, as Figure 7 confirms.

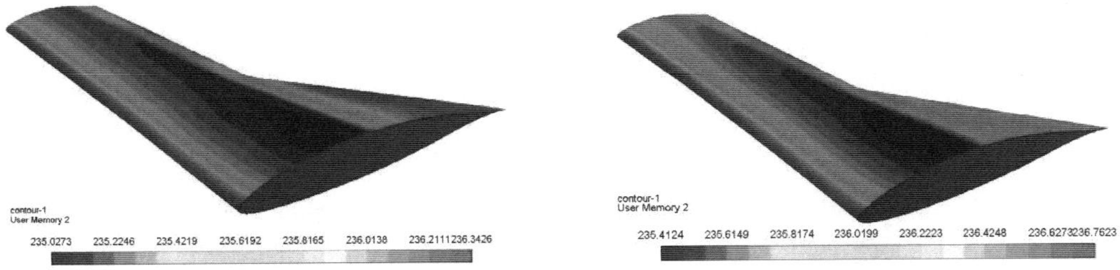

(a) Full fuel state (b) Empty fuel state

Figure 8. Frost temperature on wing surfaces at 3600s under both conditions

Figure 8 shows higher frost surface temperatures on the wing with empty fuel tanks. The trailing edge shows equal temperatures in both cases due to its greater distance from the fuel tank, which reduces the tank's thermal influence.

5. Conclusion

This study uses a frost growth prediction model to simulate the air heating conditions in aircraft fuel tanks under both empty and full fuel states before takeoff. The simulation obtains the upper and lower skin temperatures, frost accumulation, condensation water, and frost layer thickness, as well as density contour distributions under different conditions.

Frost accumulation on wings can severely compromise aircraft aerodynamics. Even small ice or frost particles of 1-2mm diameter per square centimeter on the upper wing surface prior to takeoff will roughen the airfoil. This induces premature airflow separation, reduces critical angle of attack, decreases maximum lift coefficient, increases drag, and significantly impairs flight handling stability.

The results show that under a full fuel state, the wing exhibits higher frost accumulation and frost layer thickness, while under an empty fuel state, the frost layer density is higher. The model can effectively predict frost layer distribution on the wing surface under different environmental conditions. Based on these predictions, targeted measures can be taken for either frost prevention or deicing.

Acknowledgements

This work was supported by Natural Science Foundation of Shandong Province (ZR2022MA075),and National Natural Science Foundation of China (U1833121).

References

[1] Bragg, M.B., Heinrich, D.C., Valarezo, W.O. and McGhee, R.J., 1994. Effect of underwing frost on a transport aircraft airfoil at flight Reynolds number. Journal of Aircraft, 31(6), pp.1372-1379.

[2] Hayashi, Y., Aoki, A., Adachi, S. and Hori, K., 1977. Study of frost properties correlating with frost formation types. Journal of heat transfer, 99(2), pp.239-245.

[3] O'neal, D.L., 1985. A review of frost formation in simple geometries. ASHRAE trans, 91(2), pp.267-282.

[4] Na, B. and Webb, R.L., 2004. New model for frost growth rate. International Journal of Heat and Mass Transfer, 47(5), pp.925-936.

[5] Lee, K.S., Kim, W.S. and Lee, T.H., 1997. A one-dimensional model for frost formation on a cold flat surface. International Journal of Heat and Mass Transfer, 40(18), pp.4359-4365.

[6] Wu, X., Ma, Q., Chu, F. and Hu, S., 2016. Phase change mass transfer model for frost growth and densification. International Journal of Heat and Mass Transfer, 96, pp.11-19.

[7] Ismail, K.A.R. and Salinas, C.S., 1999. Modeling of frost formation over parallel cold

plates. International Journal of Refrigeration, 22(5), pp.425-441.
[8] Chen, Y., Lu, P., Shen, C. and Zhang, Q., 2015. Experimental study on frost formation on a cold surface in low atmospheric pressure. Applied Thermal Engineering, 90, pp.86-93.

Journal of Physics: Conference Series **3056** (2025) 012010 doi:10.1088/1742-6596/3056/1/012010

Application Analysis of Temperature Prediction in UAV Equipment Compartment Based on Thermal Network Method

Yankun Chen[1], Weijian Chen[1*][0000-0001-6776-4492], Guangya Zhu[1], Chaowu Xiang[2]

[1] Key Laboratory of Aircraft Environment Control and Life Support, MIIT, Nanjing University of Aeronautics & Astronautics, Nanjing, P.R. China, 210016

[2] The Organization Department of the Qingyang District Committee of the Communist Party of China, Chengdu, P.R. China, 610031

*Corresponding author's e-mail: weijian_chen@nuaa.edu.cn

Abstract. Thermal network method is employed to predict the temperature distribution of equipment within a UAV compartment. Through node discretization, construction of thermal network diagram, and simultaneous solution of nodal equations, temperature values at each node in equipment are obtained and systematically analyzed. The results are experimentally validated, demonstrating the feasibility and accuracy of applying thermal network method to temperature prediction in UAV equipment compartments. Compared with CFD simulations, this method significantly reduces computational time requirements, suggesting its preferential adoption during preliminary thermal design phases of UAV compartments.

1. Introduction

Thermal dissipation has emerged as one of the most urgent challenges across modern engineering systems, including satellites, propulsion systems, and electronic devices. Determining optimal heat dissipation strategies and quantifying required heat removal necessitate precise temperature predictions at critical components, which serve as a critical prerequisite for thermal design. The resolution of complex thermal challenges fundamentally relies on rigorous analysis of heat transfer phenomena under diverse operational conditions[1]. Current methodologies for heat transfer analysis are primarily addressed through the finite difference method (FDM), finite volume method (FVM), and finite element method (FEM). The finite volume method (FVM) has become the most widely adopted approach in numerical heat transfer analysis, predominantly implemented through computational fluid dynamics (CFD) software. Its primary strength lies in its high computational accuracy, enabling precise simulation of complex heat transfer mechanisms. Additionally, the method exhibits exceptional geometric adaptability, allowing it to model structures with intricate geometries and address diverse engineering challenges. However, FVM is constrained by significant computational overhead, particularly in large-scale applications. Complex engineering problems impose prohibitive demands on computational time, disk storage, and memory allocation. Furthermore, even minor modifications to the computational model necessitate time-intensive computational grid regeneration and re-simulation, which substantially hinders rapid iterative problem-solving[2].

As a thermoelectric analogy-based analytical methodology, the thermal network method demonstrates operational simplicity and topological flexibility, enabling effective thermal analysis ranging from elementary heat transfer scenarios to intricate complex problems. Validated across diverse engineering disciplines, this methodology has been extensively implemented in aerospace vehicle thermal control systems, architectural energy optimization frameworks, and high-density electronics thermal management solutions.

This study applies the thermal network method to predict temperature distributions within the thermal environment of a UAV equipment compartment. Through systematic division of thermal network nodes and establishment of nodal thermal balance equations, programmed computational solutions are implemented to obtain temperature values at critical equipment nodes. The numerical results are subsequently validated against experimental measurements, demonstrating satisfactory agreement in nodal temperature predictions.

2. Introduction of thermal network method

The thermal network method operates as a thermoelectric analogy technique that utilizes electrical circuit principles, specifically Kirchhoff's Current Law (KCL) and Kirchhoff's Voltage Law (KVL), to systematically formulate equilibrium equations for complex heat transfer systems. This methodology defines temperature as the primary unknown variable and computationally solves the governing equations to determine temperature distributions and transient behaviors across structural nodes[3]. Within the thermoelectric analogy framework, heat flux corresponds to electric current, temperature difference equates to voltage potential, thermal resistance parallels electrical resistance, and thermal capacitance mirrors electrical capacitance[4]. The method divides the analyzed system into interconnected volume elements, each represented as a node, with heat transfer between nodes modeled through equivalent thermal resistances regardless of heat exchange mechanisms, thereby constructing a comprehensive thermal network. The derived nodal thermal balance equations may exhibit linear or nonlinear characteristics depending on system complexity, where problems involving substantial radiative heat transfer typically result in nonlinear formulations[5]. Such nonlinear systems are conventionally resolved using numerical approaches such as the Newton iteration method or its enhanced variants[6].

The governing thermal network equation for a specified node in steady-state analysis is:

$$\sum_{j=1}^{s} D_{ji}\left(T_j - T_i\right) + \sum_{j=1}^{r} R_{ji}\sigma\left(T_j^4 - T_i^4\right) + Q_i = 0 \quad (1)$$

In the equation, D_{ji}-thermal transfer coefficient between j and i (W/K), R_{ji}-radiation coefficient between j and i (m^2), Q_i-internal heat source of the node i (W), T-temperature (°C).

The governing thermal network equation for a specified node in transient analysis is:

$$(GC)_i \frac{dT_i}{dt} = \sum_{j=1}^{s} D_{ji}\left(T_j - T_i\right) + \sum_{j=1}^{r} R_{ji}\sigma\left(T_j^4 - T_i^4\right) + Q_i \quad (2)$$

$$D_{ji} = h_i + \frac{\lambda_i}{\sigma_i} \quad (3)$$

$$R_{ji} = \frac{1}{\dfrac{1-\varepsilon_i}{\varepsilon_i A_i} + \dfrac{1}{A_i F_{i\to j}} + \dfrac{1-\varepsilon_j}{\varepsilon_j A_j}} \quad (4)$$

In the equation, $(GC)_i$-heat capacity of the node i (J/K), ε_i-emissivity of node i, $F_{i\to j}$-view factor

3. Physical model

The UAV equipment compartment model analyzed in this study, as illustrated in Figure 1, comprises an external carbon fiber shell, six internal electronic units (four of which feature heat-dissipating fins on their upper surfaces), two U-shaped carbon fiber support brackets, and an aluminum mounting plate. The six electronic units are sequentially designated as Device 1 to Device 6 in a right-to-left spatial configuration. The geometric parameters of each component are systematically tabulated in Table 1.

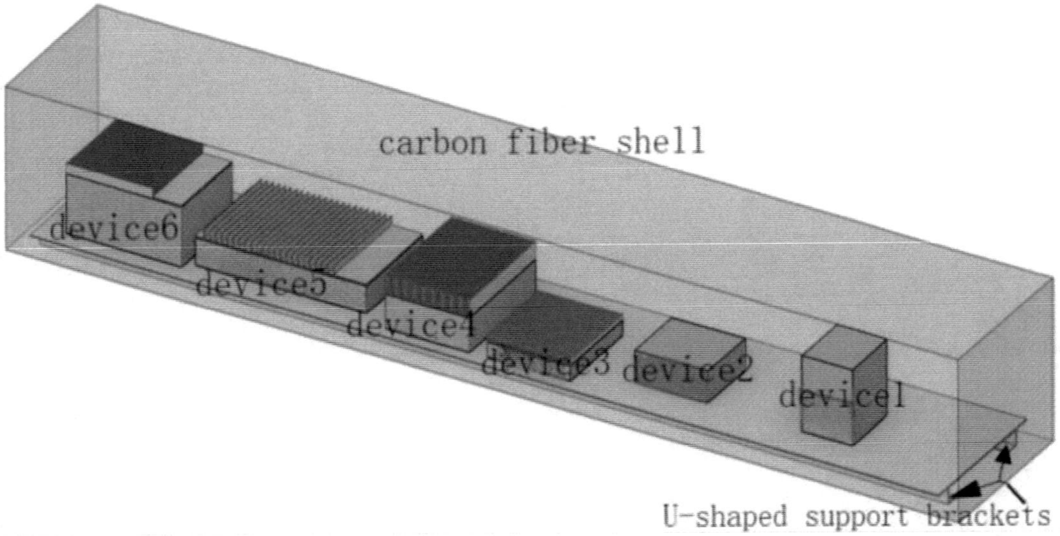

Figure 1. Schematic diagram of the equipment compartment model

4. Thermal network model

4.1. Assumed condition

The application of the thermal network method to predict temperature distributions within the UAV equipment compartment necessitates the following assumptions[7-12]:

1) The thermal and fluid dynamic influences of other UAV components on the equipment compartment are neglected in this analysis.

2) The infrared emissivity of internal components within the equipment compartment remains spectrally constant across their respective operational wavebands.

3) A uniform gas temperature distribution is assumed around individual devices, with no thermal coupling between adjacent device boundary layers.

4) A complete mutual absorption condition is postulated for infrared radiation exchange between opposing surfaces within the compartment.

4.2. Nodal network partitioning and nodal network equations

Taking the equipment compartment as the research object, a nodal network discretization is implemented under the aforementioned assumptions. This study specifically addresses a two-dimensional nodal configuration, with the spatial distribution of nodes graphically presented in Figure 2.

Table 1. Model parameter table

Model name	Length (mm)	Width (mm)	Height (mm)	Wall thickness (mm)	Quantity	Spacing (mm)	Power (W)
Carbon fiber shell	1640	280	280	1	1		
Device 1	105	87	130	3	1		20
Device 2	124	111	54	3	1		35
Device 3	155	144	26	3	1		15
Device 4	200	145	59	3	1		40
Device 5	290	170	47	3	1		100
Device 6	200	150	102	3	1		100
Fins on Device 3	155	1	8		27	4.5	
Fins on Device 4	200	1	28		37	3	
Fins on Device 5	170	1	15		23	9	
Fins on Device 6	150	1	15		51	1.98	
Aluminium alloy plate	1560	260	2		1		
U-shaped carbon fiber support bracket	1560	25	28.6	1.2	2		

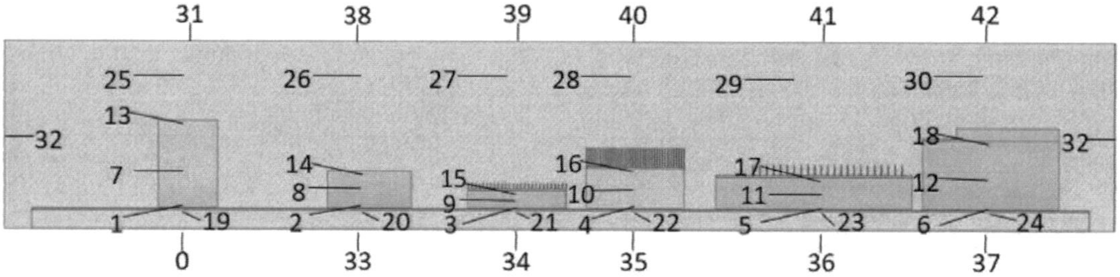

Figure 2. Equipment compartment node distribution diagram

The node numbers and their corresponding coordinates are tabulated in Table 2.

According to the discretized nodal configuration, the thermal network model is established in accordance with the schematic representation provided in Figure 3.

In the thermal network model, R_{i-j} denotes the thermal resistance between nodes *i* and *j*, while ⊗ represents the heat source applied at the node.

Table 2. Node number and corresponding positions

Node number	Corresponding positions and name	Symbol of temperature	Node number	Corresponding positions and name	Symbol of temperature
0	Undersurface temperature 1 of the equipment compartment	T_0	22	Temperature of aluminum plate under Device 4	T_{22}
1	Lower section temperature of the Device 1	T_1	23	Temperature of aluminum plate under Device 5	T_{23}
2	Lower section temperature of the Device 2	T_2	24	Temperature of aluminum plate under Device 6	T_{24}
3	Lower section temperature of the Device 3	T_3	25	Temperature of air around Device 1	T_{25}
4	Lower section temperature of the Device 4	T_4	26	Temperature of air around Device 2	T_{26}
5	Lower section temperature of the Device 5	T_5	27	Temperature of air around Device 3	T_{27}
6	Lower section temperature of the Device 6	T_6	28	Temperature of air around Device 4	T_{28}
7	Mid-section temperature of the Device 1	T_7	29	Temperature of air around Device 5	T_{29}
8	Mid-section temperature of the Device 2	T_8	30	Temperature of air around Device 6	T_{30}
9	Mid-section temperature of the Device 3	T_9	31	Upper surface temperature1 of equipment compartment	T_{31}
10	Mid-section temperature of the Device 4	T_{10}	32	Lateral surface temperature of the equipment compartment	T_{32}
11	Mid-section temperature of the Device 5	T_{11}	33	Undersurface temperature 2 of the equipment compartment	T_{33}
12	Mid-section temperature of the Device 6	T_{12}	34	Undersurface temperature 3 of the equipment compartment	T_{34}
13	Upper section temperature of the Device 1	T_{13}	35	Undersurface temperature 4 of the equipment compartment	T_{35}
14	Upper section temperature of the Device 2	T_{14}	36	Undersurface temperature 5 of the equipment compartment	T_{36}
15	Upper section temperature of the Device 3	T_{15}	37	Undersurface temperature 6 of the equipment compartment	T_{37}
16	Upper section temperature of the Device 4	T_{16}	38	Upper surface temperature2 of equipment compartment	T_{38}
17	Upper section temperature of the Device 5	T_{17}	39	Upper surface temperature3 of equipment compartment	T_{39}
18	Upper section temperature of the Device 6	T_{18}	40	Upper surface temperature4 of equipment compartment	T_{40}
19	Temperature of aluminum plate under Device 1	T_{19}	41	Upper surface temperature5 of equipment compartment	T_{41}
20	Temperature of aluminum plate under Device 2	T_{20}	42	Upper surface temperature6 of equipment compartment	T_{42}
21	Temperature of aluminum plate under Device 3	T_{21}	a	Ambient temperature	T_a

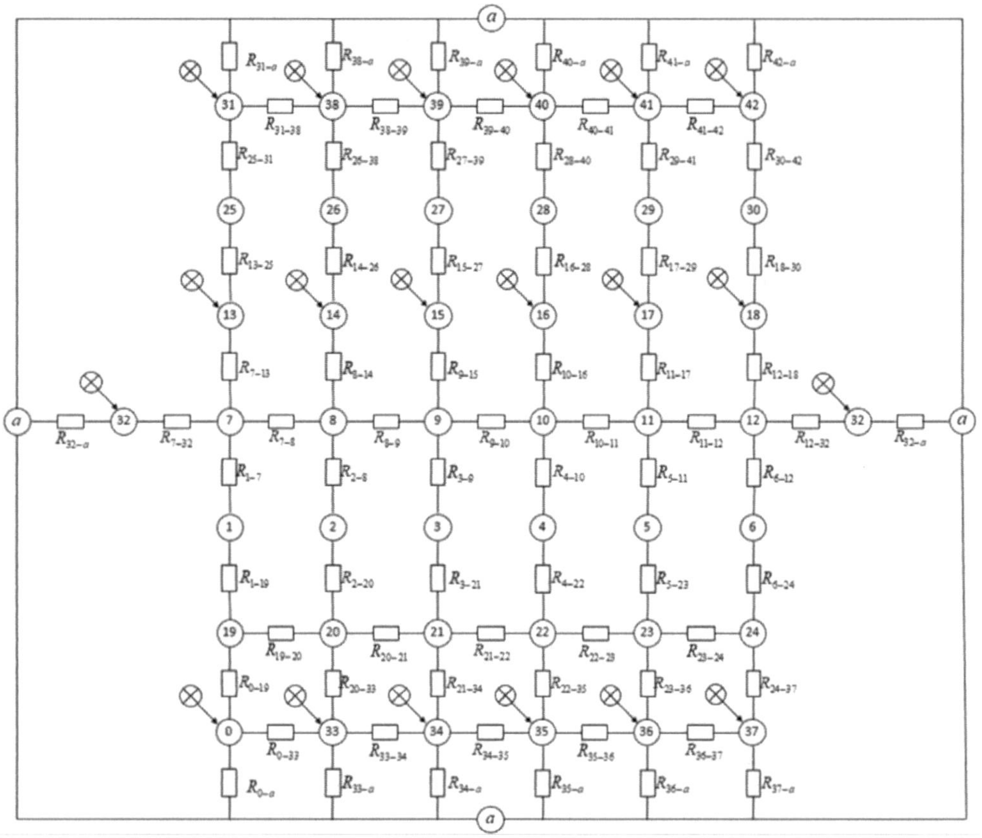

Figure 3. Thermal network diagram of the equipment compartment

The current analysis solves the nodal network equations under steady-state conditions with an ambient temperature of 40°C and ground operating conditions. The compartment air experiences no forced convection, with only natural convection driven by temperature differentials present. The external airflow velocity simulates 8 m/s corresponding to low-speed UAV flight conditions. Heat source locations are assumed at the upper sidewalls within each equipment unit. When the equipment cabin reaches thermal equilibrium, the thermal energy entering any node equals that exiting the node. Following this conservation principle, 42 thermal equilibrium equations have been formulated for the discretized 42-node system, constituting the complete thermal balance equation set for the equipment cabin. Representative thermal equilibrium equations are presented as follows:

$$
\begin{aligned}
&S_0 + \frac{T_{19}-T_0}{R_{0-19}} + \frac{T_{33}-T_0}{R_{0-33}} + \frac{T_a-T_0}{R_{0-a}} + A_0\sigma\varepsilon(T_a^4 - T_0^4) = 0 \\
&\frac{T_7-T_1}{R_{1-7}} + \frac{T_{19}-T_1}{R_{1-19}} = 0 \\
&\frac{T_8-T_2}{R_{2-8}} + \frac{T_{20}-T_2}{R_{2-20}} = 0 \\
&\cdots\cdots
\end{aligned}
\tag{5}
$$

The thermal resistance associated with conductive heat transfer can be determined using Equation (6), while the convective heat transfer resistance is calculated through Equations (7) and (8). The radiative thermal resistance, corresponding to heat transfer via radiation, is derived from Equation (9)[13].

$$R_{\lambda} = \frac{\delta}{\lambda \cdot A} \tag{6}$$

In equation, δ-wall thickness (m), λ-thermal conductivity of solid materials (W/(m·K)), A-thermal conductivity area of a solid (m^2).

$$R_{h} = \frac{1}{h \cdot A} \tag{7}$$

$$h = \frac{Nu \cdot \lambda}{l} \tag{8}$$

In equation, h-convective heat transfer coefficient (W/(m^2·K)), A-convective heat transfer area (m^2). Nu can be determined by relevant empirical formulas.

$$R_{\varepsilon} = \frac{1}{\varepsilon\sigma\left(T_i^3 + T_i^2 T_j + T_i T_j^2 + T_j^3\right)A} \tag{9}$$

In equation, ε-infrared emissivity of materials, σ-Stefan-Boltzmann constant.

The aforementioned system of equations constitutes a quartic multivariate nonlinear system. Current methodologies for solving nonlinear equation systems have been well-established, with applicable algorithms including the Newton-Raphson method and quasi-Newton methods. In this study, Python programming is employed to implement the computational solution, specifically utilizing the fsolve function from the SciPy library. This function operates as a robust solver for nonlinear equations, employing advanced numerical optimization techniques to efficiently compute solutions for nonlinear equations or equation systems. The implemented algorithm demonstrated effective resolution of the thermal equilibrium equations governing the equipment compartment through systematic numerical iteration and Jacobian matrix approximation.

4.3. Results

The nodal temperatures of the device calculated using the thermal network method are presented in Table 3.

Table 3. Calculated node temperature on the Device

Device	Node number			Node temperature(K)		
Device 1	1	7	13	332.57	338.06	347.20
Device 2	2	8	14	348.57	351.86	355.88
Device 3	3	9	15	340.40	341.27	342.22
Device 4	4	10	16	349.54	352.28	355.54
Device 5	5	11	17	363.37	371.13	383.27
Device 6	6	12	18	368.53	382.66	396.13

The calculation results reveal significant temperature variations among the devices due to differences in heat dissipation power. Devices 5 and 6, with heating powers reaching 100W, exhibit operating temperatures substantially higher than other devices. In contrast, Device 3 demonstrates the lowest maximum temperature at the upper node among all six devices, attributable to its minimal heat generation and enhanced heat transfer through upper fins. The thermal power of Device 4 is slightly higher than that of Device 2. However, the fins on the top surface of Device 4 significantly increase the heat exchange area with ambient air, thereby enhancing convective heat transfer. This results in a marginally lower maximum temperature at the top of Device 4 compared to Device 2, demonstrating the substantial benefits of fin structures in improving heat dissipation. Furthermore, the vertical temperature gradient of each device increases proportionally with thermal power output. For Devices 2-4, their relatively low thermal power generation and compact vertical dimensions result in reduced thermal resistance along the longitudinal conduction path, maintaining temperature differences between upper and lower nodes below 10K.In contrast, Device 1 exhibits a larger temperature gradient despite its moderate 20W thermal power, primarily due to its taller structure which introduces greater thermal resistance along the vertical conduction path, leading to significant heat loss during downward transmission. For high-power Devices 5-6, their thermal generation rate exceeds the heat dissipation capacity, causing temperature escalation beyond 100°C. The thin wall thickness creates substantial thermal resistance for downward heat transfer, forcing most heat flux to be dissipated through upper fins via convective cooling. Consequently, these devices demonstrate extreme vertical temperature gradients reaching 20°C between upper and lower nodes.

To validate the reliability of the thermal network method in predicting temperature distributions within UAV equipment compartment, experimental verification will be conducted in the following section.

5. Experimental verification

5.1. Experimental device and process

The experimental system and model are illustrated in Figures 4 and 5. This investigation employ an equivalent analogy method using internal tube flow heat transfer to simulate external convective heat transfer. The equipment compartment under study features a square cross-section, resulting in identical dimensions for heat exchange channels on all four lateral surfaces. Based on these dimensional parameters, the internal heat exchange channel is designed as a rectangular passage measuring 0.287 m in width and 0.04 m in height. The physical prototype of this configuration is presented in Figure 6.

To simulate the self-generated heat of electronic devices, polyimide electric heating films are attached to the upper internal surfaces of six devices. The output power is regulated through a stabilized voltage supply to emulate their thermal generation. Thermocouples are strategically positioned at the central region of the external surfaces of each devices and other critical measurement points. These thermocouples are connected to a data acquisition system, where specialized data collection software is employed to continuously record temperature variation data from both the devices and key monitoring locations. Concurrently, thermal insulation cotton is applied to the exterior surfaces of all connecting pipelines and heat exchange channels to minimize thermal interference with experimental outcomes.

During the experiment, the equipment compartment model is positioned inside the environmental simulation chamber. The laboratory gas supply system delivers inflow air under ground atmospheric pressure at 40°C. Upon stabilization of relevant parameters, the regulated power supply is activated to energize the electrical heating film while simultaneously initiating data acquisition and recording. Following thermal stabilization of all simulated equipment, the gas supply and power delivery are terminated, data collection is halted, and the experimental procedure is completed.

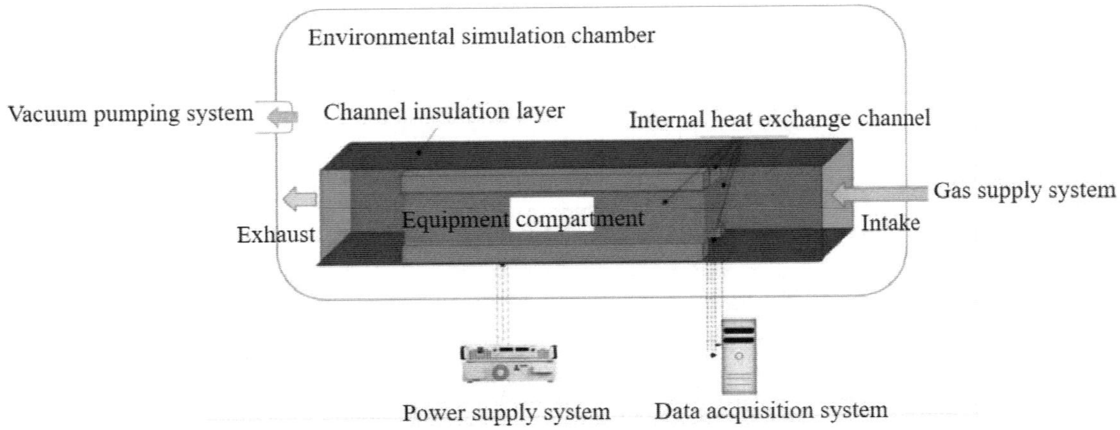

Figure 4. Experimental system diagram

Figure 5. Experimental model

Figure 6. Internal heat exchange channel

5.2. Comparison of experimental results

The experimental results are compared with the calculated intermediate node temperatures from the thermal network method, as presented in Table 4. It is observed that the temperature distributions of six devices obtained through the thermal network method exhibited strong agreement with those derived from experimental measurements, with the minimum temperature occurring at Device 1 and the maximum temperature at Device 6. Comparative analysis reveal maximum absolute and relative errors of 14.02 K and 3.66%, respectively, between the two methodologies, demonstrating satisfactory consistency that validates the feasibility and accuracy of the thermal network approach.

However, certain discrepancies are identified, primarily attributable to model simplifications inherent in the lumped parameter methodology. The nodal discretization strategy, which substitutes continuous domains with discrete nodes, inevitably impacts computational precision through mesh density dependence. Additional error sources include approximations in thermal resistance calculations, particularly the reliance on empirical formulations for convective thermal resistance and simplified treatments of radiative heat transfer. Furthermore, cumulative errors may arise from fundamental assumptions underpinning the thermal network methodology prior to implementation.

Table 4. Comparison of temperature between two methods

Device	Thermal network method(K)	Experimental method(K)	Absolute error	Relative error (%)
Device 1	338.06	335.5	2.56	0.76
Device 2	351.86	348.85	3.01	0.85
Device 3	341.27	332.07	9.2	2.69
Device 4	352.28	342.85	9.43	2.68
Device 5	371.13	358.28	12.85	3.46
Device 6	382.66	368.64	14.02	3.66

6. Conclusion

This study employs the Thermal Network Method to calculate the temperatures of components within the UAV equipment compartment under operational conditions of 40°C ambient temperature and ground atmospheric pressure. The computational results are compared with experimental data, revealing a maximum absolute error of 14.02 K and a maximum relative error of 3.66%. The good agreement between numerical and experimental results validates the feasibility and accuracy of the thermal network method for thermal analysis in equipment compartments. For subsequent UAV equipment compartment designs, the thermal network method can be effectively implemented during the preliminary thermal design phase. This approach enables rapid and accurate acquisition of temperature distribution patterns for internal components, thereby providing valuable references for thermal control system development. Compared with CFD methodologies, this technique eliminates the requirement for complex and time-consuming meshing procedures, significantly reducing computational preparation time while maintaining satisfactory engineering accuracy.

Acknowledgments

This research was funded by National Natural Science Foundation of China (No.52106247)

References

[1] Kang Q, Li S W and Guo J L 2006 The thermal network method outline Industrial Heating vol 35
[2] Xiang C W 2024 Research on Thermal Ervironment and Thermal Control Technology of Solar UAV Equipment Cabin

[3] Qu J X 2004 Summary of thermal analysis of spacecraft systems Infrared p 20-27
[4] Wu Y X, Huang Y Z and Deng E P 2022 Survey of thermal network modeling for power devices Electrical Technology
[5] Weng J H and Zhang T Y 2016 Thermal network method and its application in thermal analysis and calculation Development & Innovation of Machinery & Electrical Products vol 29
[6] Li Q Y, Mo Z Z and Qi L Q 1992 Numerical Solution of Nonlinear Equations (Beijing: Science Press)
[7] Yang X 2014 Temperature Calculation of Submersible Motor Based on Thermal Network Method
[8] Yang P 2013 The Thermal Analysis of Helicopter Main Reducer Based on Thermal Network Method
[9] Song J H 2023 Thermal Modeling and Optimization of Three-dimensional Integrated Circuits Based on Finite Element and Equivalent Thermal Network Methods
[10] Jia Z J 2004 Investigation on Establishing a Numerical Method for the Thermal Analysis of a Satellite
[11] Huang F 2011 The Thermal Analysis of Epicyclic Gear Reducer Based on Thermal Network Method
[12] Chen W 2014 Research of Thermal Network Method on the Spaceborne Electronic Equipment
[13] Zhang J Z and Chang H P 2015 Heat Transfer (Beijing: Science Press)

ANALYSIS OF CONTACT THERMAL RESISTANCE IN SPACE REMOTE SENSING INSTRUMENT: THEORETICAL AND EXPERIMENTAL STUDY

Fu-long Liu*, Feng Yu,Chang-chun GAO, Chang LIU， Guang-lang WEI
Thermal Control Department
Institute of Space Mechanics & Electronics, Beijing
Mailbox 5142-230, Beijing 100094, China
Tel:+86-15001323169, Fax:+86-010-68113603, E-mail:66173514@qq.com

ABSTRACT

The contact thermal resistance between solid surfaces cannot achieve the observed contact area at the macroscopic scale; actual contact at the microscopic level only consists of discrete points or small areas, This paper investigates the coupled heat transfer between the refrigerator and heat pipe in a spacecraft refrigeration system. A disk computational model was established, and through thermal simulation calculations and vacuum test experiments, the contact heat transfer coefficients for dry contact, GD414 silicone rubber filler, thermal insulation pads, and thermal silicone grease were obtained. This model can be used to predict thermal coupling results in engineering applications and effectively guide engineering practices.

KEYWORDS

contact thermal resistance, disk computational model, thermal simulation calculations,vacuum test experiments

INTRODUCTION

With the development of aerospace thermal control technology, solid conduction remains the primary method for heat dissipation from high heat flux sources, such as core CMOS devices, refrigerators, management controllers, cooling controllers, and secondary power supplies in satellite sensors. The heat generated by these devices during operation must be transferred to the satellite's external radiator surfaces through high-conductivity components like groove heat pipes or loop heat pipes. This process inevitably involves contact between solid surfaces, resulting in thermal resistance and temperature differences.

Traditional thermal design methods rely on empirical parameters input into simulation models for thermal calculations. However, as satellite and sensor integration increases, there is a shortage of thermal control resources such as radiator surfaces and power consumption, rendering traditional empirical parameters inadequate. Thus, minimizing and optimizing temperature differences caused by contact thermal resistance has become critical to the success of thermal control designs.

This paper, grounded in practical applications, measures the thermal resistance and temperature differences during the solid-to-solid contact between heat pipes and refrigerator hot ends under vacuum conditions. Based on these measurements, it proposes a numerical calculation model for solid-solid contact thermal resistance. Using this model, the research predicts changes in contact thermal resistance when adding heat-conductive materials such as silicone rubber, indium foil, and thermal silicone grease.

THERMAL RESISTANCE MODEL

When solid surfaces come into contact, the actual contact area observable at the macroscopic scale is not achieved. At the microscopic level, contact occurs only at discrete points or small areas [1-3]. Due to the microscopic unevenness of the two contacting surfaces, the contact is incomplete, leading to contact thermal resistance. The development of contact thermal resistance models primarily focuses on the surface topography characteristics and deformation issues. In this paper, we refer to the disc calculation model proposed by Cooper in 1969, further, the following formula is proposed:

$$R = \frac{1}{4a\lambda_m} u(a/b) \tag{1}$$

$$u(a/b) \approx (1 - a/b)^{1.5} \tag{2}$$

$$1/\lambda_m = 1/\lambda_1 + 1/\lambda_2 \tag{3}$$

λ_1 means thermal conductivity of structure 1 ,W/(m·K)
λ_2 means thermal conductivity of structure 2 ,W/(m·K)
a means contact radius, mm
b means disk computational radius, mm
R means contact thermal resistance, $(m^2 \cdot K)/W$

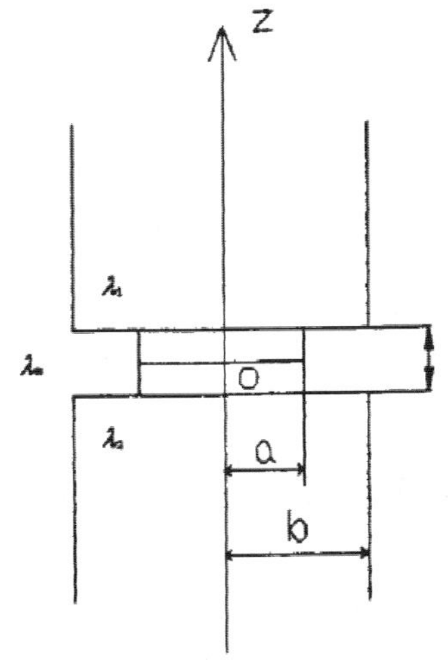

Figure 1 disk computational model

Before establishing the contact thermal resistance model, the following assumptions are made:

1. The thermal physical properties at the contact interface are assumed to be isotropic and constant;
2. The heat flux completely penetrates the thermal conduction process, meaning all the heat fully passes through the contact interface without any loss;
3. The micro-deformations occurring at the contact surface follow an elastic model.

Journal of Physics: Conference Series **3056** (2025) 012011 doi:10.1088/1742-6596/3056/1/012011

As shown in the contact thermal resistance test model in Figure 2, the heat flux enters from the backside of the refrigeration machine cold plate and exits axially along the heat pipe side, while the remaining surfaces are considered to be in an adiabatic state. According to the definition of thermal resistance, the contact thermal resistance can be expressed as:

$$R_t = \Delta t_c / q \tag{4}$$

Δt_c means contact surface temperature difference, K

q means heat flux density, W/ m2

The temperature distribution of the refrigeration cold plate and heat pipe is solved using the numerical method, Δt_c is calculated, and then substituted into equation (4) to obtain R_t.

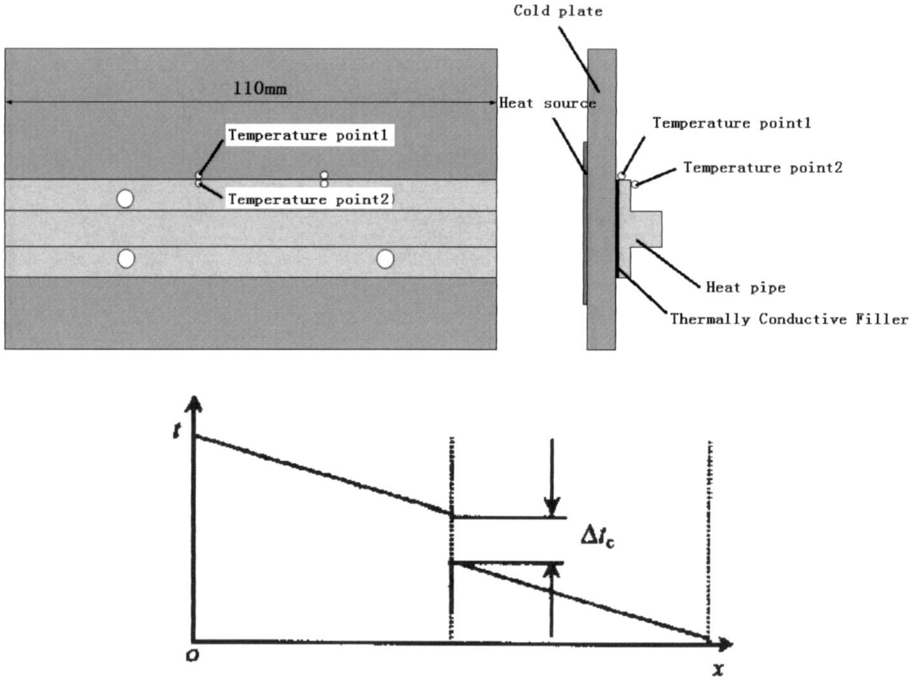

Figure 2 thermal resistance model

Since the contact surface adopts a disk calculation model, a two-dimensional axisymmetric temperature distribution is obtained. Laplace equation for constant thermal conductivity, without internal heat sources, and in steady-state conditions is established in cylindrical coordinates as follows [5]:

$$\frac{\partial^2 t}{\partial z^2} + \frac{\partial^2 t}{\partial r^2} + \frac{1}{r}\frac{\partial t}{\partial r} = 0 \tag{5}$$

The thermal boundary condition of the second kind specifies the given heat flux (q):

$$q = -\lambda \left(\frac{\partial t}{\partial n} \right)_w \tag{6}$$

n means normal direction of the cross-section

z means the line of symmetry

Journal of Physics: Conference Series **3056** (2025) 012011 doi:10.1088/1742-6596/3056/1/012011

r means the radius

The following is the functional expression for solving the equation [6]:

$$J\left[t(z,r)\right] = \iint_D \frac{\lambda r}{2}\left[\left(\frac{\partial t}{\partial z}\right)^2 + \left(\frac{\partial t}{\partial r}\right)^2\right]dzdr + \oint_\Gamma qtrds \quad (7)$$

Discretize and grid the temperature field, solve for the vertex temperatures t_i, t_j, t_m within the control volume, and convert the functional into a problem of solving a multivariate function for the temperatures at control points.

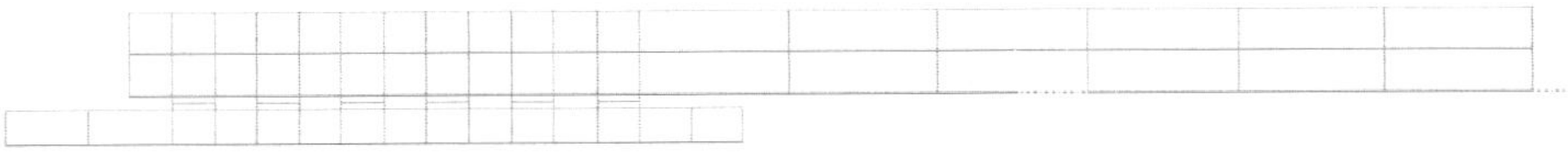

Figure 3 grid partition diagram

By solving the temperature difference between the two contacting surfaces, the contact thermal resistance can be calculated.

SIMULATION AND EXPERIMENTAL RESULTS

Using this model, thermal analysis results were calculated for different contact states between the chiller cold plate and heat pipe under vacuum conditions, and compared with the measured data inside the vacuum chamber.

Table 1 Vacuum measured data

	Power-Q	nominal contact area	Screw tightening torque-N	Temperature difference	Contact heat transfer coefficient
Direct contact	13.43	0.0033	1	1.11	3666
Filled Silicone Rubber	13.43	0.0033	1	0.79	5152
Filled Thermal Insulation Pad	13.43	0.0033	1	0.63	6460
Thermal silicone grease application	13.43	0.0033	1	0.22	18499

Table 2 Thermal Simulation Results

	Thermal conductivity	Temperature difference	Contact heat transfer coefficient
Vacuum	0	1.14	3570
Silicone Rubber	0.3	0.78	5218
Thermal Insulation Pad	3.5	0.61	6672
Thermal silicone grease (Doped with Metal Powder)	18.8	0.22	18499

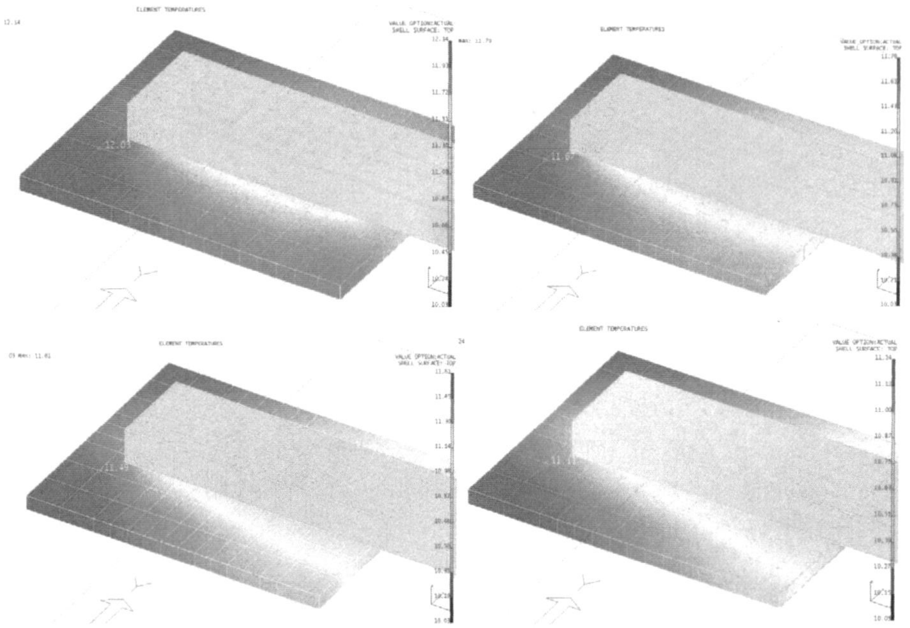

Figure 4 Temperature Field Distribution

The heat transfer process on solid surfaces under vacuum conditions can be equivalent to conductive heat transfer through the actual contact area between two surfaces and radiative heat transfer through the nominal contact area between the two surfaces. Adding thermal conductive fillers can expand the heat transfer pathways. As shown in the analysis results from Table 2, the contact heat transfer coefficient has been improved to a certain extent after adding thermal conductive fillers, and it is positively correlated with the thermal conductivity of the fillers themselves[7-9].

CONCLUSION

In this study, a disk calculation model is employed to compute the contact heat transfer coefficient between solid surfaces and simulate the heat transfer process between the cold plate of a refrigerator and heat pipes. The thermal simulation results have achieved good agreement with vacuum measurement results, indicating that this model possesses high applicability in engineering applications. Additionally, during the practical application of the remote sensor, thermal conductive fillers can effectively enhance heat exchange between solid surfaces. The thermal conductivity of these fillers is positively correlated with the contact heat transfer coefficient.

REFERENCES

[1]. Wang Liang-an， Zhao Jian-feng. Review on the Prediction of Contact Thermal Resistance [J]. Chinese Science,2011,41(5): 545-556.
[2]. Berry M.V， Lewis Z.W. On the Weierstrass-Mandelbrot Fractal Function. Proc. R. Soc. Lond 1980,A370:459-484
[3]. Gong Zhao， Yang Chun-xin. The Simplification of Contact Thermal Resistance

Theoretical Models [J]. Journal of Engineering Thermophysics,2007.9, 28(5): 850-852.
[4]. Cooper.M.G，Mikic.B.B，Yvanovich.M.M. Thermal contact conductance. Lnt.J. heat mass transfer 1969 12:279
[5]. Yang Shi-ming，Heat Transfer (Second Edition) [M]. Beijing：Higher Education Press，1987.66
[6]. Tao Wen-quan，Computational Heat Transfer (Second Edition) [M]. Beijing：Xi'an Jiaotong University Press，2005.71-80
[7]. Zhao Lan-ping，Xu Lie. Theoretical Analysis of Thermal Contact Resistance Between Solid Interfaces[J]. Chinese Space Science And Technology, 2003,23(4) : 6-11.
[8]. Zhang Tao，Xu Lie，Xiong Wei，Zhao Lan-ping. Comparison and Analysis of Theoretical Models in Contact Thermal Resistance [J]. low-temperature physics and superconductivity,1998.5,26(2):58-64.
[9]. Cao Jun，Hu Pei-jun，Ying Ji. Finite Element Analysis of Spindle Thermal Characteristics Based on Contact Thermal Resistance [J]. Journal of Electromechanical Engineering, 2008,2.25(2):20-22.

Journal of Physics: Conference Series **3056** (2025) 012012 doi:10.1088/1742-6596/3056/1/012012

Numerical and Experimental Study of the Phase Change Materials Heat Exchanger Coupled with Fluid Loops in Spacecraft Thermal Control

Wang Yan*, Fu Zhendong, Wang Yuying, Chen Ling, Wu Xianlin, Jin Yu, Meng Fankong, Huang Jinyin, Yu Xingang, Cao Jianfeng, Zhang Hongxing, Miao Jianyin

Beijing Institute of Spacecraft System Engineering, National Key Laboratory of Spacecraft Thermal Control, Beijing, 100094, P.R. China

Corresponding author's e-mail: brucewanglegend@foxmail.com

Abstract: To address the complex thermal environment in lunar orbit, this paper presents a plate-fin phase change materials heat exchanger (PCMHE) coupled with a fluid loop. By utilizing the phase change materials (PCMs) for heat storage during melting and heat release during solidification, the heat dissipation capacity of the spacecraft can be dynamically adjusted under varying external thermal conditions. A simplified simulation model was established for simulation analysis, calculating the heat transfer performance, melting/solidification characteristics. The conclusion drawn from simulation analysis is that 22kg of PCMs can control the fluid temperature within 8℃. A performance test rig was constructed to validate the design. Experiment results demonstrate that the outlet fluid temperature remained within 8°C under the given inlet fluid temperature and flow rate. The proposed fluid loop coupled with the PCMHE provides an effective thermal control solution for manned spacecraft in lunar orbit.

1 Introduction

In the past decade, plenty of research has been conducted on the application of Phase Change Materials (PCMs) in spacecraft thermal control[1~3]. PCMs have been widely utilized in spacecraft thermal control systems, particularly for managing temperature of electronic components. Applicability of PCMs technology in space-related missions is promising because its' capacity of dissipating large heat flux levels and maintaining surface temperature within certain limits. Challenges are also posed because of PCMs' low thermal conductivity. It is of vital importance to explore methods

Journal of Physics: Conference Series **3056** (2025) 012012 doi:10.1088/1742-6596/3056/1/012012

to enhance their thermal performance, such as incorporating high thermal conductivity filler materials like metallic foams and carbon fibers[4].

The significant fluctuations in external heat flux in lunar orbit pose challenges to spacecraft thermal management. Integrating PCMs into fluid loop systems offers a viable solution. PCMs exhibit high latent heat capacity, enabling efficient heat storage/release within minimal temperature variations, making them ideal for spacecraft thermal control[5]. This study designs a plate-fin PCMHE directly coupled with a fluid loop to regulate heat dissipation by storing excess heat during the sunlit zone and releasing it during the shadow zone on the lunar orbit.

Plate-fin heat exchangers, known for compactness and high efficiency, are widely used in aerospace thermal control systems[6~8]. For instance, China's Shenzhou spacecraft employs plate-fin heat exchangers for thermal regulation[9]. The U.S. Orion spacecraft also utilizes plate-fin PCMHEs filled with hydrocarbon-based PCMs to enhance heat dissipation in lunar orbit[10][11]. While PCM-based thermal control has been applied in orbit, no fluid loop-coupled PCMHEs have been implemented domestically. This paper fills this gap by designing, simulating, and experimentally validating a PCMHE tailored for lunar orbit thermal environments.

2 PCMHE Design

The PCMHE adopts a plate-fin structure (Figure. 1). To enhance heat transfer, classic plate-fin structure is adopted in the PCMHE, comprising alternating layers of PCMs layer and heat transfer fluid (HTF) channels. PCMs Layer and HTF channel are connected by partition plates.

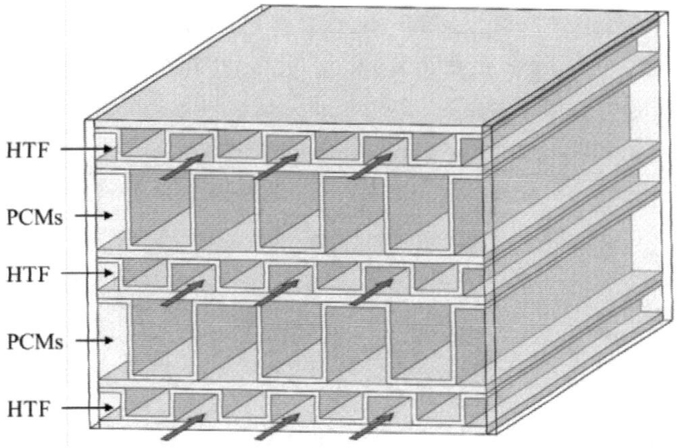

Figure 1. Diagram of the plate-fin PCMHE

Under the lunar orbit, the significant difference in lunar infrared radiation between the shadow and sunlit regions results in a substantial variation in heat dissipation capability. Therefore, this paper proposes integrating a phase change heat exchanger into the spacecraft's thermal control fluid loop to regulate the spacecraft's heat dissipation capability under different external thermal environments by coupling with the fluid loop. The phase change heat exchanger is located at the radiator outlet, and the fluid loop schematic is shown in Figure 2. One side of the phase change heat exchanger contains the thermal control fluid loop working medium, while the other side contains the phase change working medium. The bidirectional heat exchange between the fluid loop working medium and the phase

Journal of Physics: Conference Series **3056** (2025) 012012 doi:10.1088/1742-6596/3056/1/012012

change working medium is achieved through a liquid coupling method.

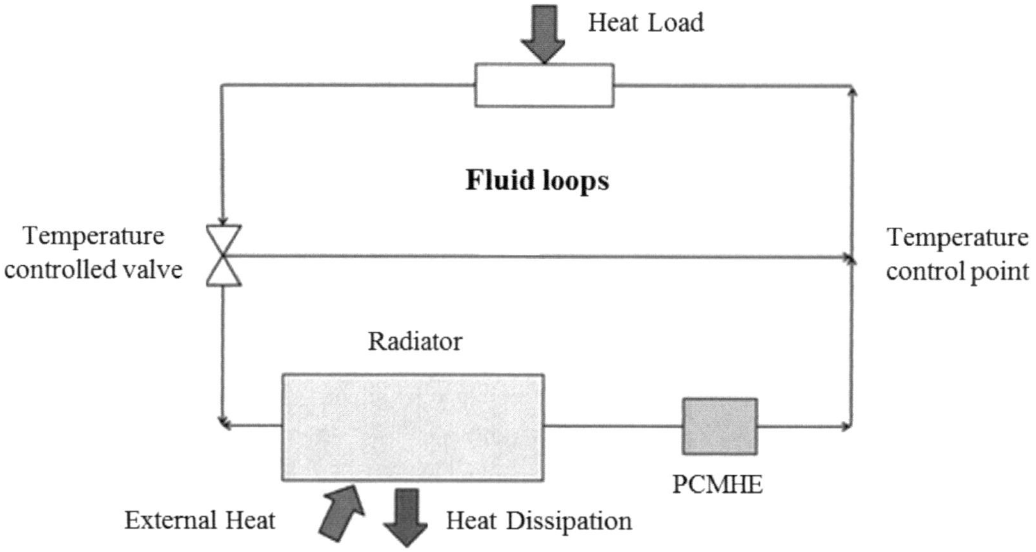

Figure 2. Diagram of the PCMHE coupled with fluid loops in spacecraft

The loop equipment and radiator are designed to maintain a fluid loop temperature of 8 °C. Assuming the entire spacecraft has a heat dissipation requirement of 2600 W and operates in a lunar orbit at an altitude of 200 km, the initial design and thermal simulation determine the fluid inlet temperature of the phase change heat exchanger in lunar orbit. The analysis of the radiator's heat dissipation capability over one orbital period under external heat flux conditions in lunar orbit is shown in Figure 3. Within one orbital period, the phase change heat exchanger needs to have a thermal storage capacity of 5000 kJ and ensure that the outlet temperature of the phase change heat exchanger is controlled within 8 °C.

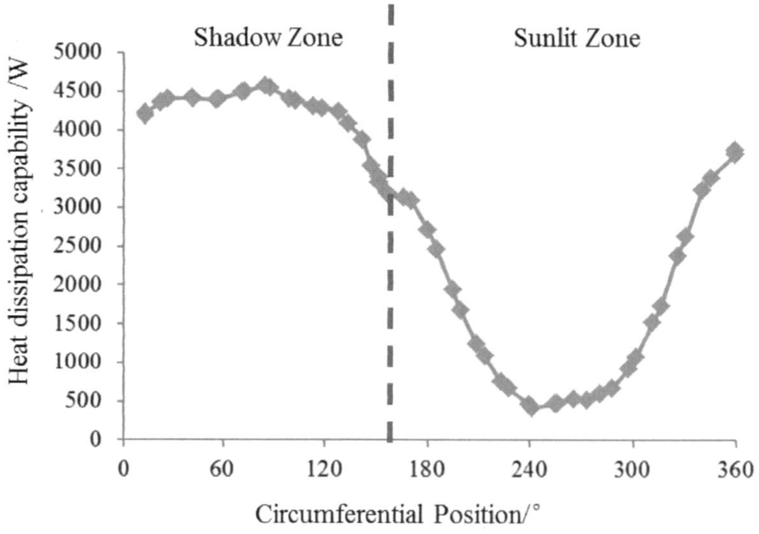

Figure 3. Simulation result of the radiator heat dispassion capacity in a lunar orbit cycle

3 Thermal Simulation

3.1 Simplified Model

To improve simulation efficiency, it is necessary to abstract and simplify the complex plate-fin phase change heat exchanger, resulting in a theoretical model that accurately reflects its physical mechanisms while being convenient for modeling and simulation calculations. Given the periodic appearance of the plate-fin structure in the product, the model is simplified to a single basic unit, as shown in Figure 4. This unit includes 2 blocks of phase change material and the adjacent flat plates and fins, as well as 1/2 of the loop 1 fluid channel and 1/2 of the loop 2 fluid channel.

The choice of 2 blocks of phase change material is primarily due to the significant differences in heat transfer between the phase change material and the loop 1 and loop 2 fluids, which result from the welding between the fins and the flat plates in the height direction. Additionally, in a single basic unit, the fins on both sides only include half the thickness of the actual fins in the product, allowing the sides to be treated as periodic boundaries for simulation. Considering that both the heat source and the cold source come from the fluids, the boundaries can be further simplified to adiabatic boundaries.

Given the symmetry in fluid flow, both the upper and lower fluid channels only include half of the actual fluid channels. The plane where the fluid contacts the flat plate is simulated with a no-slip boundary condition, while the middle plane of the fluid channel (the boundary of the fluid channel in the basic unit model that does not contact the flat plate) can be simulated with a slip boundary condition. The flow rate within the fluid channel is calculated based on the actual flow area. The length of the model in the direction of fluid flow is determined by the total mass of the designed phase change material.

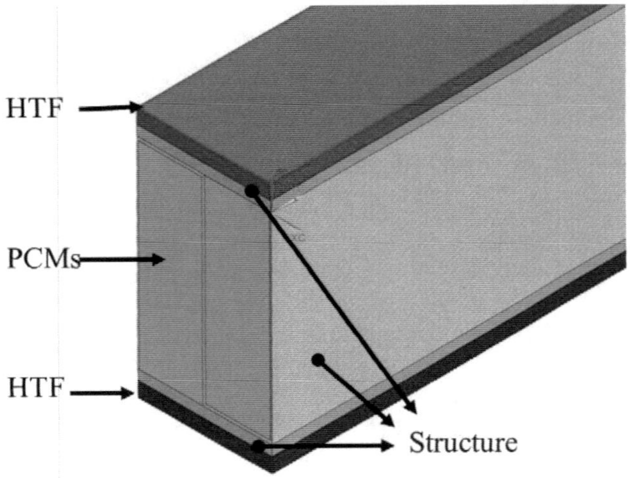

Figure 4. Simplified model of basic unit in PCMHE

The mentioned model aligns with the thermodynamic boundaries of a single chamber within the actual phase change heat exchanger, but it deviates from the chambers located at the edges of the heat exchanger. In spacecraft applications, the outer surface of the phase change heat exchanger is covered with multiple layers of thermal insulation, which can be approximated as adiabatic during operation. Additionally, the number of units in contact with the boundaries is relatively small compared to the

Journal of Physics: Conference Series **3056** (2025) 012012 doi:10.1088/1742-6596/3056/1/012012

total number of units. Therefore, the thermodynamic performance of a single basic unit model can be used to estimate the actual heat transfer characteristics of the entire phase change heat exchanger.

3.2 Simulation setup

In this paper, the alkane phase change material n-tetradecane is selected. The material properties used in the simulation are listed in Table 1.

Table 1 Thermal properties of n-tetradecane

Parameter	Value
density/ ($kg \cdot m^{-3}$)	825
conductivity/ ($W \cdot m^{-1} \cdot K^{-1}$)	0.15
heat capacity/ ($J \cdot kg^{-1} \cdot K^{-1}$)	1560 (Solid) /2070 (Liquid)
phase change temperature/K	278.7
Latent heat/ ($kJ \cdot kg^{-1}$)	228

In the simulation analysis, the phase change heat exchangers are arranged in series, and the inlet fluid temperature is specified based on the operating conditions. The calculation process is as follows:

1. Simulate the first phase change heat exchanger by setting the inlet temperature and obtain the outlet temperature of the first heat exchanger.
2. Simulate the second phase change heat exchanger using the outlet temperature of the first heat exchanger as the inlet temperature for the second heat exchanger, and obtain the outlet temperature of the second heat exchanger.
3. Continue this process for subsequent phase change heat exchangers, with the outlet temperature of each previous heat exchanger serving as the inlet temperature for the next one.
4. The fluid flow rate remains constant for each phase change heat exchanger.

The initial temperatures of both the fluid and the phase change material are 5.5°C, and the phase change material is in a solid state. The analysis focuses on the outlet fluid temperature and the melting status of the phase change material.

3.3 Simulation Results

When conducting simulation analysis of the phase change heat exchanger, the calculations are performed using a determined fluid inlet temperature. The simulation analyzes the fluid outlet temperature of the phase change heat exchanger over two orbital periods in lunar orbit under different amounts of phase change material. The results are shown in Figure 5. When the phase change material in the phase change heat exchanger exceeds 22 kg, it can ensure that the fluid outlet temperature does not exceed 8 °C, and the thermal storage capacity of a single-phase change heat exchanger exceeds 2500 kJ.

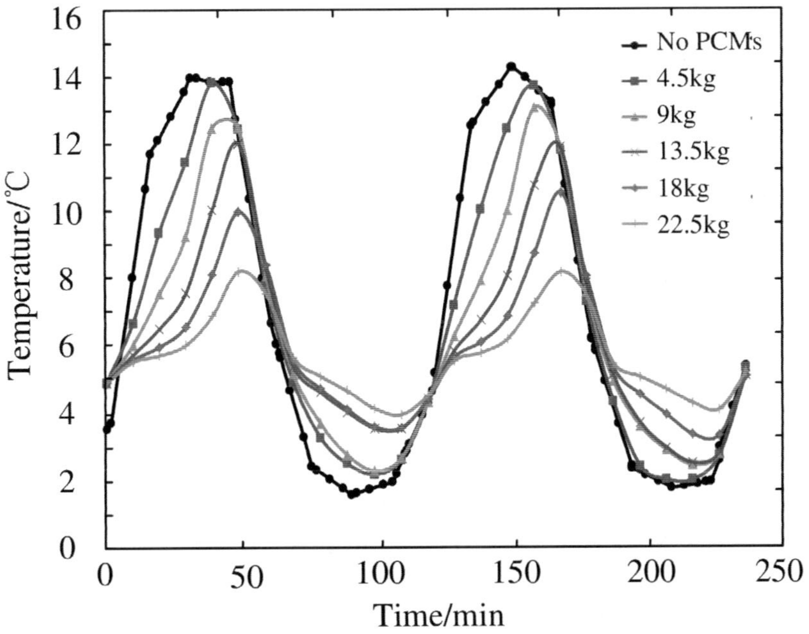

Figure 5. Simulation results of PCMHE fluid outlet temperature with various amount of PCM

Based on the simulation results, using 22 kg of phase change material, the phase change heat exchanger is coupled with the fluid loop for thermal analysis. The calculations consider the temperature control valve's ability to regulate the fluid flow through the phase change heat exchanger, resulting in the variation of the phase change heat exchanger's inlet temperature over time, as shown in Figure 6. After the product is manufactured, the single-unit product can be tested under the following temperature conditions.

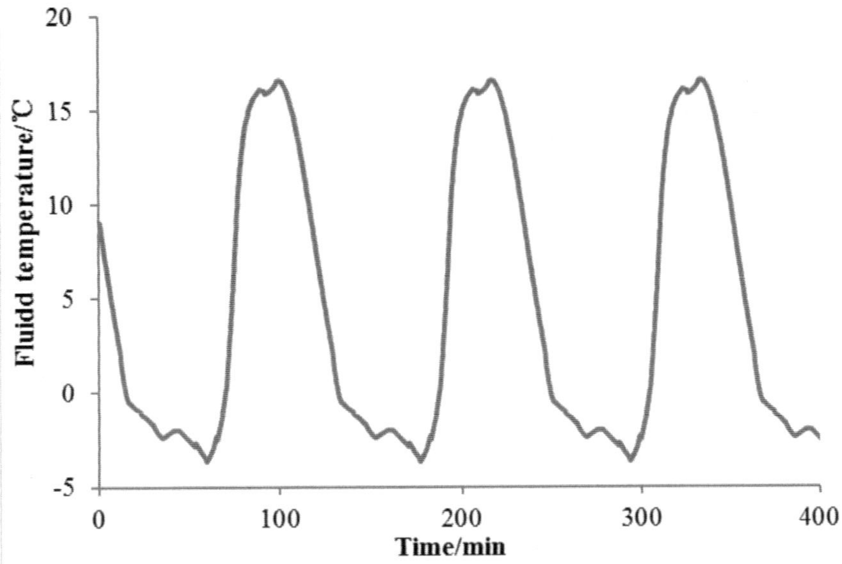

Figure 6. Simulation results of the PCMHE fluid inlet temperature working in the fluid loops

Journal of Physics: Conference Series **3056** (2025) 012012 doi:10.1088/1742-6596/3056/1/012012

4 Experiment validation

4.1 Experiment design

A prototype of the phase change heat exchanger was developed according to the design in this paper, and a loop test bench was set up to conduct performance tests. The phase change heat exchanger performance testing system, as shown in Figure 7a, mainly includes the following components:

1. Simulated Fluid Loop: A loop to simulate the fluid flow through the heat exchanger.
2. Mechanical Pump Group: Pumps to circulate the fluid through the system.
3. Temperature, Pressure, and Flow Measurement System: Instruments to measure the temperature, pressure, and flow rate of the fluid.
4. Heat Loading System: Equipment to apply and control the heat load on the system.
5. Data Acquisition and Processing System: Devices and software to collect and process the test data.

This setup allows for comprehensive testing and evaluation of the phase change heat exchanger's performance. A prototype PCMHE and test rig (Figure 7b) were developed.

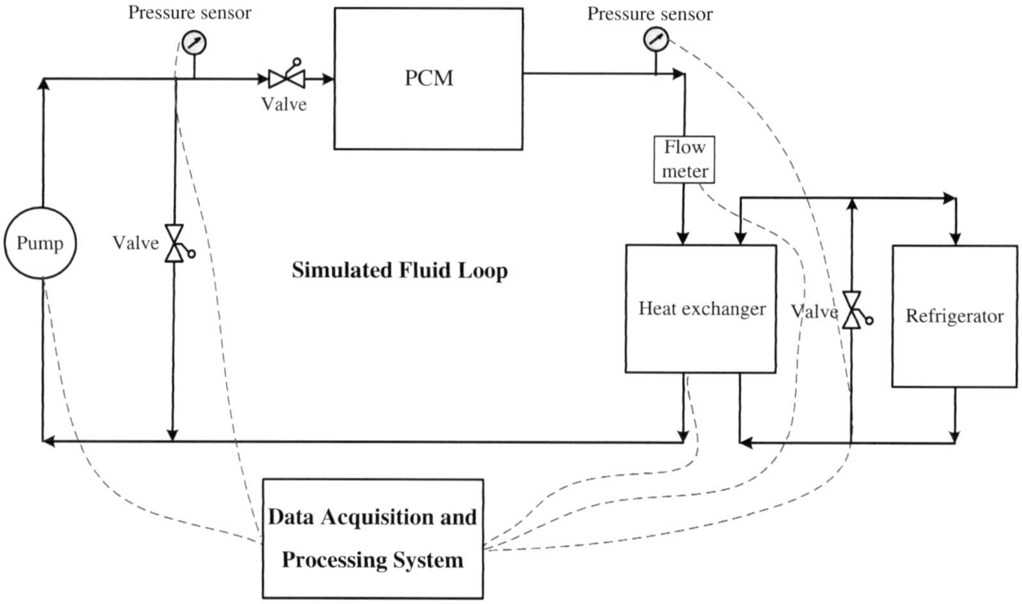

(a) PCMHE Performance Test Schematic Diagram

（b）PCMHE Performance Test Facility

Figure 7. PCMHE performance test system

4.2 *Experiment results*

Using the experimental system described in the previous section, the experimental verification was conducted during the melting and condensation cycles of the phase change heat exchanger's working fluid. The temperature changes and distribution of the fluid side inlet and outlet of the phase change heat exchanger are shown in Figure 8. The phase change material in the phase change heat exchanger can completely melt and solidify within the predetermined time. Under the given fluid inlet temperature conditions, the outlet temperature remains below 8°C, meeting the temperature performance requirements.

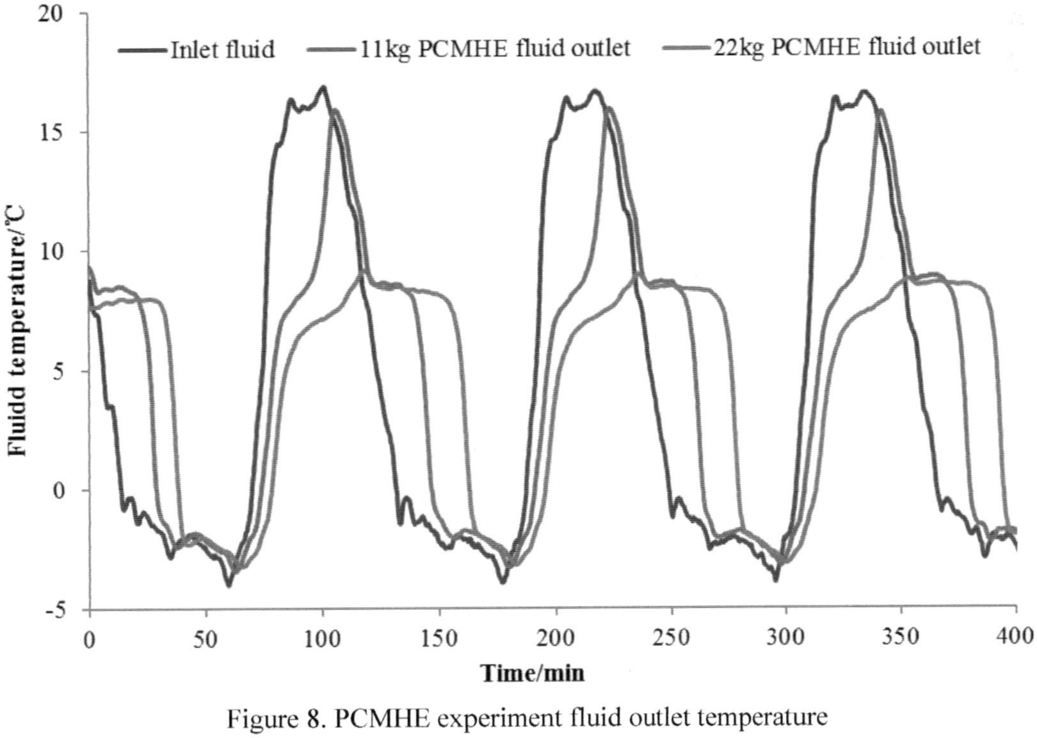

Figure 8. PCMHE experiment fluid outlet temperature

5 Conclusion

This paper designs a plate-fin phase change heat exchanger coupled with a fluid loop. Through simulation analysis, the internal heat transfer characteristics are studied, and the amount of phase change material and the fluid temperature conditions for experimental verification are determined. The experimental system is then set up to validate the rationality of the phase change heat exchanger design and to verify the satisfaction of the temperature performance requirements. The fluid loop scheme coupled with the phase change heat exchanger proposed in this paper can serve as an effective thermal control method for manned spacecraft in lunar orbit.

Acknowledgments

This research is supported by National Natural Science Foundation of China: 52202506, Beijing Nova Program (CN): Z211100002121078 and 20240484646.

References

[1] MICHAEL K C. Using paraffin PCM to make optical communication type of payloads thermally self-sufficient for operation in Orion crew module [C]// 14th International Energy Conversion Engineering Conference. Salt Lake City, UT. American Institute of Aeronautics and Astronautics, 2016: 1-7.

[2] PAN Aigang, WANG Junbiao, ZHANG Xianjie. Application status and prospect of phase change temperature control technology in aerospace thermal control field [J]. Materials Review, 2013, 27(12): 113-119 (in Chinese).

[3] MAN Yahui. Research on latent heat mechanism of phase transformation and its application technology [D]. Changsha: National University of Defense Technology, 2010 (in Chinese).

[4] JANICKI M, NAPIERALSKI A. Modelling electronic circuit radiation cooling using analytical thermal model [J]. Microelectronics Journal, 2000, 31(2000): 781-785.

[5] WANG Songhan. Plate-fin heat exchanger [M]. Beijing: Chemical Industry Press, 1984 (in Chinese).

[6] B.S.V. Prasad, S.M.K.A. Gurukul. Differential Method for Sizing Multi-stream Plate Fin Heat Exchangers [J]. Cryogenics, 1987, 27: 257-262.

[7] YU Jiangzu. Theory and design of the heat exchanger [M]. Beijing: Beijing University of Aeronautics and Astronautics Press, 2005 (in Chinese).

[8] CAO Wenxue, SHI Qian, BIAN Jiang, et al. Simulation of heat transfer characteristics of plate-fin heat exchanger based on co-simulation [J]. Journal of China Petroleum University (Natural Science Edition), 2022, 44(3): 148-154 (in Chinese).

[9] LI Jun. Research on performance of airborne multi-strand fluid-fin heat exchanger [D]. Nanjing: Nanjing University of Aeronautics and Astronautics, 2015 (in Chinese).

[10] Wang X Y, Yuko J R. Orion active thermal control system dynamic modeling using Simulink/MATLAB [C]// Proceedings of the 48th AIAA Aerospace Sciences Meeting. Orlando, Florida. American Institute of Aeronautics and Astronautics, 2010: 1-19.

[11] WANG Xiao-Yen. Independent verification and validation of the Orion multi-purpose crew vehicle active thermal control system performance [C]// NASA report. 2012: 1-10.

Journal of Physics: Conference Series **3056** (2025) 012013 doi:10.1088/1742-6596/3056/1/012013

Thermal Control Design and In-Orbit Verification of the Star Sensors on Taijing-3 01 Satellite

Yinxue Yan*, Jiong Zhang, Yue Zhang, Pan Lu, Qingle Kong

Beijing Institute of Space Mechanics&Electricity, No.104 Youyi Road, Haidian Beijing, China

*Corresponding author's e-mail: yanyinxue@126.com

Abstract. To address the complex thermal environment challenges faced by satellite star sensors during in-orbit operation, this study developed a composite thermal control scheme that combines passive heat dissipation and active temperature control. The thermal control objective is accomplished by optimizing the radiator layout based on orbital external heat flux analysis, employing multilayer insulation and heater circuit design. Simulation results and in-orbit data show that under extreme thermal operating conditions, this scheme can still maintain the temperature of the star sensor within the temperature control accuracy range, effectively verifying the rationality of the design.

1. Introduction

The star sensor is a core device for satellites to achieve high-precision attitude measurement, and its performance directly affects the positioning and stability control capabilities of spacecraft [1]. The Taijing-3 01 satellite operates in a sun-synchronous orbit, and the star sensors on it will encounter complex thermal environments such as periodic solar radiation, Earth albedo, and infrared radiation during in-orbit operation [2,3]. Unstable temperature fluctuations can cause deformation of optical components and reduce circuit performance. Therefore, it is crucial to conduct precise thermal control design for star sensors [4].

Currently, most star sensor thermal control systems combine passive insulation and active heating [5]. Reference [6] conducted a thermal design for a micro star sensor module and found that the scheme using graphite as the heat conduction path is reasonable; Reference [7] proposed a thermal control design scheme for a radiation small-cabin star sensor; Reference [8] proposed a method that combines heat pipes, thermal control coatings, multi-layers, and electric heaters. This study combines the characteristics of the Taijing-3 01 satellite, and through external orbital heat flux analysis, thermal control scheme design, and simulation verification, ensures the temperature stability of the star sensors under extreme thermal operating conditions. By analyzing the in-orbit temperature data of the star sensors during normal operation, the correctness of the thermal control system design and its in-orbit adaptability are verified.

2. Overview of Star Sensors and Temperature Control Requirements

2.1. *Composition of Star Sensors*

The star sensors on the Taijing-3 01 satellite consist of three parts: Star Sensor A, Star Sensor B, and Star Sensor C, which are installed on the -Y, +Y, and +X sides of the satellite respectively. The entrance

Journal of Physics: Conference Series **3056** (2025) 012013 doi:10.1088/1742-6596/3056/1/012013

aperture diameters are 100mm and 91.6mm, as shown in Figure 1. Their core components include sunshades and circuit boxes, with heat dissipations of 0.8W (Star Sensor A) and 1.2W (Star Sensor B and Star Sensor C), as shown in Figure 2.

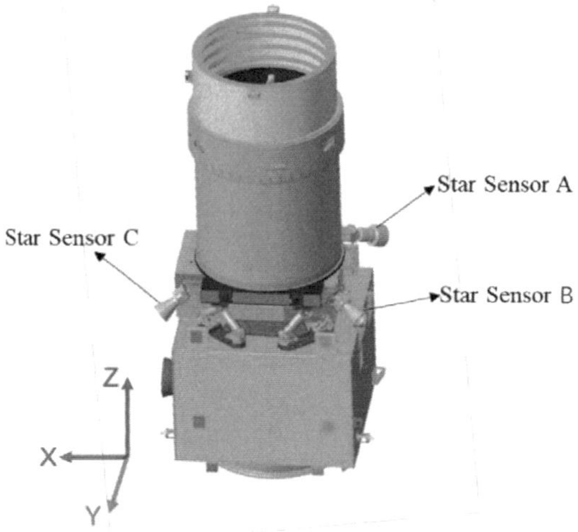

Figure 1. Installation Layout of Star Sensors

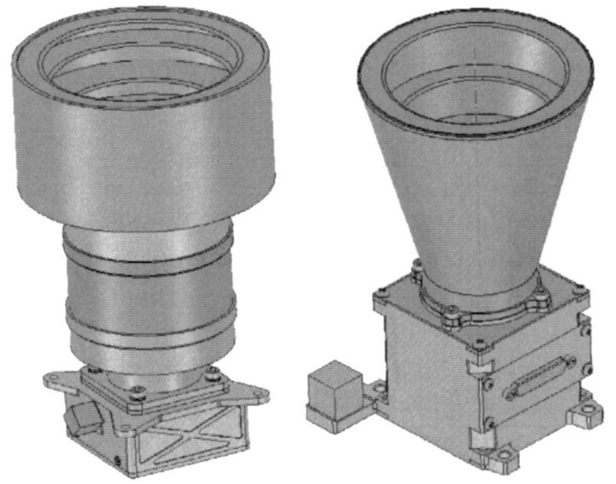

(a)Star Sensor A (b) Star Sensor B,C

Figure 2. Star Sensor Models

2.2. Temperature Control Indexes of Star Sensors

The star sensors need to meet the following temperature requirements:

Table 1. Temperature Control Requirements of Star Sensors

	Heat Dissipation(W)	Temperature(℃)
Star Sensor A	0.8	-40~+40 (0~+10 is optimal)
Star Sensor B	1.2	-20~+40 (20±2 is optimal)
Star Sensor C	1.2	-20~+40 (20±2 is optimal)

3. Thermal Control System Design of Star Sensors

3.1. Orbital External Heat Flux Analysis

Through orbital external heat flux calculations, the +Y side of the satellite is a long-term shaded side with the lowest absorbed heat flux (0~46.8 W/m^2 at sun-pointing cruise mode), which is suitable as a heat dissipation radiator. The outer surfaces of the star sensors are all covered with multi-layers, so the heat flux at the entrance apertures of the star sensors is mainly considered. The extreme heat flux values at the inlet of the star sensor under different angles between the solar vector and the orbital plane (β angles) and various operational modes of the camera are presented in Table 2.

Table 2. Extreme Values of Heat Flux at Star Sensors

Item		Working Condition	Heat Flux Maximum Condition (W/m^2)		Heat Flux Minimum Condition (W/m^2)	
Entrance Apertures	Star Sensor C	Working Condition1	Imaging Mode, β Angle Maximum, Roll -30°		Data Transmission Mode, β Angle Maximum	
			Peak Value	1294	Peak Value	1018.5
			Average Value	630	Average Value	305.1
	Star Sensor B	Working Condition 2	Imaging Mode,β Angle Minimum, Roll -30°		Data Transmission Mode, β Angle Maximum	
			Peak Value	1108	Peak Value	299.2
			Average Value	518	Average Value	163.6
	Star Sensor A	Working Condition 3	Imaging Mode,β Angle Maximum, Roll 10°		Data Transmission Mode, β Angle Maximum	
			Peak Value	1440	Peak Value	1282.7
			Average Value	744	Average Value	454.7
Heat dissipation radiator		Working Condition 4	Imaging Mode, β Angle Minimum, Roll 30°		Sun-pointing cruise mode， β Angle Maximum	
			Average Value	25	Average Value	17.6

3.2. Thermal Control System Design

3.2.1 Heat dissipation radiator Design

Two heat dissipation radiators are opened towards the +Y side of the satellite, which are used to dissipate heat for the three star sensors respectively. Among them, Star Sensor B and Star Sensor C share a heat dissipation radiator, and Star Sensor A has a separate heat dissipation radiator. Each star sensor is connected to the heat dissipation radiator through two 18×φ10mm aluminum-ammonia axial groove heat pipes. The heat dissipation radiator is installed on the main load-bearing frame of the camera through a bracket for insulation. The heat pipes of the -Y-side star sensor are also fixed on the main load-bearing frame of the camera through a bracket, as shown in Figure 3.

Journal of Physics: Conference Series **3056** (2025) 012013 doi:10.1088/1742-6596/3056/1/012013

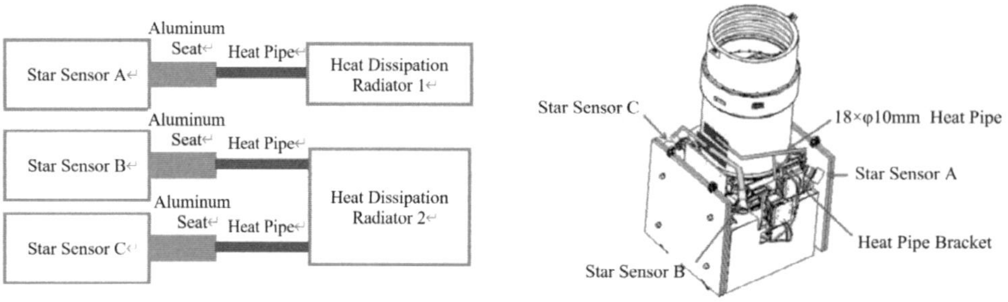

Figure 3. Schematic Diagram of the Star Sensor Heat Dissipation Scheme

The material of the heat dissipation radiator is 2A12O aluminum, and the surface of the heat dissipation radiator uses an OSR thermal control coating. The dimensions of the heat dissipation radiator are shown in Table 3. The outer surface of the sunshade of Star Sensor A is pasted with an F46 film for heat dissipation, and the local surface of the sunshade of Star Sensor B and C are pasted with an F46 film, as shown in Figure 4(The red area is covered with an F46 film, and the green and yellow areas are covered with multi-layers).

Table 3. Dimensions of Heat dissipation radiators

	Area (mm^2)	Thickness (mm)
Heat dissipation radiator 1	19500	3
Heat dissipation radiator 2	19600	3

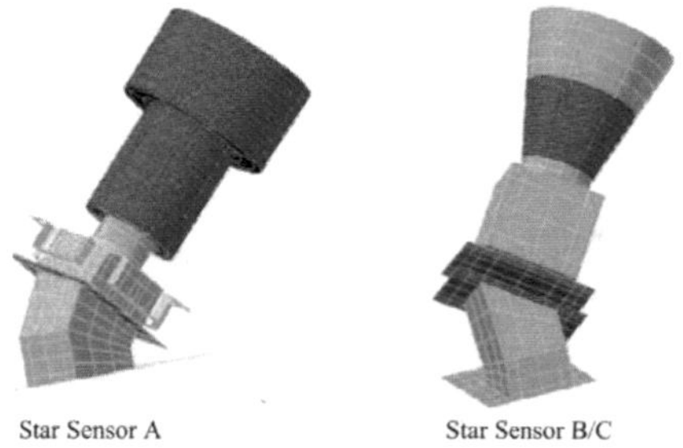

Figure 4. Heat Dissipation and Multi-layer Design of Star Sensor Sunshades

3.2.2 Insulation Design
The star sensors are installed with insulation between the star sensors and the star sensor brackets through 10mm polyimide insulation gaskets (outer diameter less than 8mm), and titanium alloy screws are used. The outer surfaces of the star sensors and the heat dissipation radiators of the star sensors, except for the +Y side, are covered with multi-layer insulation components.

3.2.3 Active Temperature Control Design
For each star sensor, a 15W temperature control circuit is set on the adapter aluminum seat of Star Sensor A, and the temperature control threshold is set as [3.7℃, 4.3℃]; a 10W temperature control circuit is set on the adapter aluminum seat of Star Sensor B and Star Sensor C respectively, and the temperature control threshold is set as [19.7℃, 20.3℃]. The total temperature control power consumption of the three star sensors is 35W.

Journal of Physics: Conference Series **3056** (2025) 012013 doi:10.1088/1742-6596/3056/1/012013

3.3. Simulation Calculation and Analysis

3.3.1 Model and Parameters

The thermal FEA Mesh of the star sensor model is established, as shown in Figure 5. Based on the orbital external heat flux calculation results in Table 2, the computational parameters for the high- and low-temperature operational scenarios are determined, as shown in Table 4. For the high-temperature case, the coating properties are set at the End of Life (EOL) [9], while for the low-temperature case, the coating properties are set at the Beginning of Life (BOL).

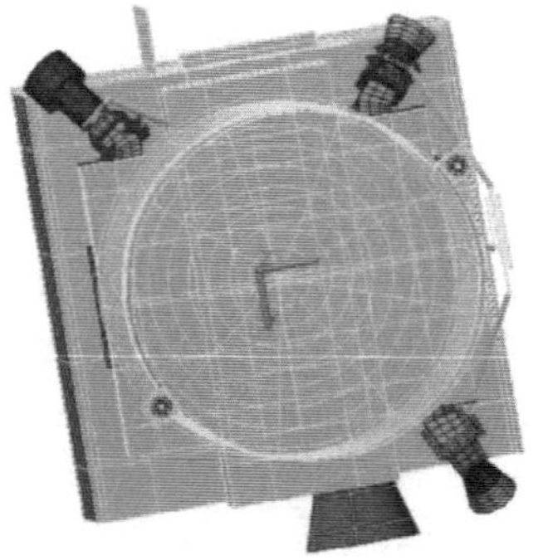

Figure 5. Thermal Analysis Model of Star Sensor

Table 4. Simulation Results of the Star Sensor Temperature

Scenario		β Angle	Working Mode	Boundary (Bracket) Temperature (℃)	Comment
High-temperature	Case1	Maximum	Imaging Mode, Roll -30°	22	Apertures of Star Sensor C Orbital Heat Flux Maximum
	Case2	Minimum	Imaging Mode, Roll -30°	22	Apertures of Star Sensor B Orbital Heat Flux Maximum
	Case3	Maximum	Imaging Mode, Roll 10°	22	Apertures of Star Sensor A Orbital Heat Flux Maximum
	Case4	Minimum	Imaging Mode, Roll 30°	22	Radiator Heat Flux Maximum
Low-temperature	Case1	Maximum	Data Transmission Mode	18	Apertures of Star Sensor A Orbital Heat Flux Minimum
	Case2	Maximum	Data Transmission Mode	18	Apertures of Star Sensor B&C Orbital Heat Flux Minimum
	Case 3	Maximum	Sun-pointing cruise mode	18	Radiator Heat Flux Minimum

3.3.2 Result Verification

Table 5 shows the simulation results, and all the indicators of the star sensors meet the design requirements.

Table 5. Simulation Results of the Star Sensor Temperature

	Star Sensor A (℃)	Star Sensor B (℃)	Star Sensor C (℃)
High-temperature Case 1	4.95~6.25	19.5~21.6	18.4~21.6
High-temperature Case 2	5.1~8.3	19.6~21.6	18.3~20.6
High-temperature Case 3	4.3~9.8	19.5~21.6	18.4~21
High-temperature Case n 4	5.7~6.9	19.5~21.6	18.4~21.8
Low-temperature Case 1	5~6.2	19.5~21.6	18.0~19.8
Low-temperature Case 2	5~6.5	19.5~21.6	18.0~19.7
Low-temperature Case 3	5~6.25	19.1~21.2	19.5~21.3
Temperature Requirements	0~10	20±2	20±2
Whether it Meets the Requirements	Yes	Yes	Yes

4. Analysis of On-Orbit Temperature Conditions

The operational temperature performance of the star sensors is presented in Figure 6. Flight data demonstrates that Star Sensor A maintains a thermal variation between 4.02°C and 4.2°C, while Star Sensors B and C exhibit a slightly wider operational range of 19.9°C to 20.2°C. These temperatures are within the temperature control threshold range and are consistent with the simulation results, verifying the effectiveness of the thermal control scheme.

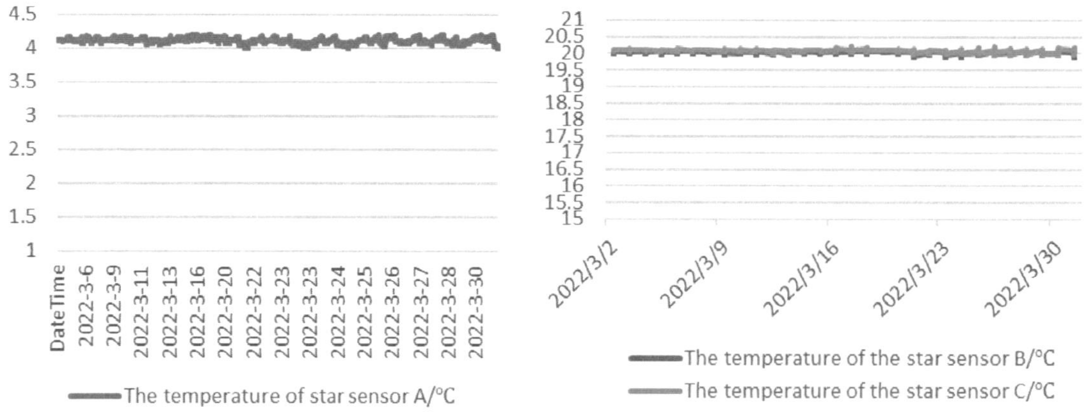

Figure 6. On-orbit Temperature of Star Sensors

5. Conclusion

1) Collaborative Thermal Control Framework

The thermal control scheme of the star sensor proposed in this study, through the coordinated design of optimizing the heat dissipation radiator, adopting multi-layer insulation, and active heating, can still ensure temperature control accuracy under extreme working conditions.

2) Verification and Engineering Value

Both the simulation results and in-orbit data have verified the feasibility of this scheme, providing an important reference for the thermal control design of star sensors of similar spacecraft.

References

[1] Janson S, Welle R. The NASA Optical Communication and Sensor Demonstration Program[C]. 27th Annual AIAA Conference on Small Satellites, USA, 2013 pp 1-7

[2] Li J L, Yan SH Z, Cai R Y. Thermal Analysis of Composite Solar Array Subjected to Space Heat Flux [J]. Aerospace Science and Technology, 2013, 27(6) pp 84-94.

[3] Zhao Z M, Wang B, Gao J. Preliminary research on the thermal design methods of the geosynchronous orbit starting camera [J]. Spacecraft Recovery & Remote Sensing, 2010, 31(3) pp 34-40

[4] Yang Q, Wang R, Jiang L F, et al. Thermal Control Design of Star Sensor for Remote Sensing Satellite[J]. Spacecraft Engineering, 2023, 32 (6) pp 74-78

[5] Zhang J H, Jia Z H, Guo L, et al. Thermal Design and Simulation Verification of High-precision Integrated Star Sensor in Near-Earth Orbit [J]. Infrared Technology, 2024,46(4) pp 400-405

[6] Lu J W, Wang L H, Su SH, et al. Thermal control design and simulation of micro star sensor[J]. Infrared and Laser Engineering, 2022, 51(11) pp 265-270

[7] Han Z W, Zhao J F, Zhao Q Q, et al. Thermal Control Design for Star Sensor on GEO Satellite[J]. Spacecraft Engineering, 2013, 22 (3) pp 47-52

[8] Zhang Y, Zhao J ZH, Zhang X, et al. Thermal design of an integrated star sensor on lunar orbit with simulation validation[J]. Spacecraft Environmental Engineering, 2016, 33(5) pp 516-520

[9] Feng Weiquan, Ding Yigang, Yan Dekui, et al. Study on long-term Degradation of Solar Absorptance Properties of Geostationary Satellite Thermal Control Coating [J] CHINESE SPACE SCIENCE AND TECHNOLOGY, 2005,25(2)pp 34-40

Journal of Physics: Conference Series **3056** (2025) 012014 doi:10.1088/1742-6596/3056/1/012014

Research on GNSS landslide deformation monitoring method based on virtual Reference station

Chen He[2],XiaoQing Wang[3]*,Xinye Dai[1],Hongzhen Yang[2],JunLi Wu[3],WeiBin Sun[1],Zhuang Peng[1]

[1] GNSS Research Center of Wuhan University,129 Luoyu Road, Hongshan District, Wuhan, Hubei Province, 430079, China

[2] State Grid Zhejiang Electric Power Co., LTD， No.219 Shimin Street, Shangcheng District,Hangzhou City, Zhejiang Province, 310016, China

[3] National Geomatics Center of China (NGCC), No.28 Lianhuachi West Road, Haidian District, Beijing, 100830, China

*Corresponding author's e-mail: xqwang@ngcc.cn

Abstract: GNSS are now routinely employed to monitor structural and geologic deformation in assets such as dams, bridges, and landslide slopes. Yet typical landslide-monitoring workflows depend on purpose-built reference stations, which considerably inflate deployment and maintenance expenses. To overcome this limitation, we present a GNSS deformation-monitoring framework that utilises a virtual reference station (VRS). The VRS is generated from observations collected by neighbouring continuously operating reference stations (CORS), removing the necessity for a local physical benchmark while still delivering centimetre-to-millimetre accuracy. The methodology was validated at roughly 30 landslide sites across low-latitude southern China, a region characterised by strong ionospheric activity. When the observation span exceeded four hours, the proposed system produced horizontal errors under 5 mm and vertical errors below 10 mm, meeting the stringent accuracy demands of landslide-warning applications. Because midday ionospheric disturbances in low-latitude zones degrade interpolation fidelity, double-difference ionospheric terms must be explicitly modelled during monitoring-end processing. Furthermore, substantial elevation gaps between the monitoring receiver and CORS antennas introduce appreciable double-difference tropospheric biases that likewise require estimation.

Keyword: VRS; Landslide; GNSS; CORS; Deformation monitoring.

1. Introduction

Owing to its all-weather operability, real-time capability and centimetre-level accuracy, GNSS has become a standard tool for deformation surveillance of dams, bridges and unstable slopes [1–4]. In 2000, researchers from the French Geoscience Laboratory and Italy's University of Padua applied GNSS to track a landslide in the Southern Alps. Even under harsh field conditions, a single-frequency static GPS solution delivered 24 mm horizontal and 74 mm vertical accuracy [5]. By 2006, the Tsing Ma Bridge (Hong Kong) had implemented a high-precision RTK-based deformation-monitoring scheme that achieved millimetre-level accuracy [6]. In 2009, the University of Bologna employed CORS observations to monitor ground deformation in northern Italy, attaining millimetre-scale accuracy in both horizontal and vertical components [7]. A 2013 collaboration between Incheon University (South Korea) and Mansura University (Egypt) deployed GPS sensors on the Mansula Bridge for deformation monitoring. By combining RTK observations with wavelet principal-component denoising, positioning accuracy improved by roughly 14–16 %, achieving millimetre-level performance [8].

The traditional GNSS landslide deformation monitoring method generally needs to build one or more GNSS reference stations near the monitoring area for each landslide geological hazard hidden danger point, which leads to the increase of the cost of the monitoring system, which is not conducive to the large-scale application and popularization of GNSS deformation monitoring technology. However, many countries / regions have established CORS services covering their whole territory. On the one hand, making full use of these existing infrastructure can greatly reduce the cost of deformation monitoring, in addition, CORS stations have a relatively stable datum, which is also conducive to the application of deformation monitoring. Therefore, a large number of scholars have proposed to use CORS reference station network for deformation monitoring [9,10,11]. Zhang Mingzhi et al. generated virtual reference stations (VRS) after fixing the ambiguity on the server end. Then, the monitoring end used the RTK dynamic and static solution mode to estimate the displacement change of the monitoring point. The results showed that centimeter-level positioning accuracy could be obtained [12]. Wang et al. improved the accuracy of the VRS monitoring method by using the net adjustment. Results reveal that, with observation sessions exceeding eight hours, planar positioning errors remain below 5 mm and vertical errors within 10 mm. Nevertheless, when the reference and monitoring stations are separated by tens of kilometres and have substantial height differences, tropospheric delays cannot be fully cancelled, thereby degrading monitoring accuracy.

In this paper, an improved GNSS landslide deformation monitoring method based on VRS is studied. To improve the tropospheric delay interpolation accuracy, the tropospheric interpolation model considering additional height correction (AHC) is adopted at the server-end data processing. The algorithm experiment and result analysis are carried out by using the real data of frequent landslide disasters in the south of China.

2. GNSS landslide deformation monitoring method based on VRS

2.1 Server-end data processing

The procedure largely mirrors the standard network-RTK workflow. The server end firstly uses the observation of the CORS to fix the double-difference ambiguities. Considering that the distance between the two CORS stations is 19-40 km, the ionosphere-free (IF) linear combinations are adopted to suppress first-order ionospheric errors [13,14]. The corresponding observation model is expressed as:

$$\begin{aligned} \nabla\Delta P^{i,j}_{\mathrm{a,b},IF} &= \nabla\Delta\rho^{i,j}_{\mathrm{a,b}} + m^{i,j}_{\mathrm{a,b}} \times \Delta T_{\mathrm{a,b}} + \varepsilon(\nabla\Delta P^{i,j}_{\mathrm{a,b},IF}) \\ \nabla\Delta L^{i,j}_{\mathrm{a,b},IF} &= \nabla\Delta\rho^{i,j}_{\mathrm{a,b}} + m^{i,j}_{\mathrm{a,b}} \times \Delta T_{\mathrm{a,b}} + \lambda_{IF} \times \nabla\Delta N^{i,j}_{\mathrm{a,b},IF} + \varepsilon(\nabla\Delta L^{i,j}_{\mathrm{a,b},IF}) \end{aligned} \tag{1}$$

Where, $\nabla\Delta$ represents double-difference operator; i,j represents satellite number, $\mathrm{a,b}$ represents station number; $\nabla\Delta P^{i,j}_{\mathrm{a,b,IF}}, \nabla\Delta L^{i,j}_{\mathrm{a,b,IF}}$ represent double-difference pseudo-range and carrier phase ionosphere-free combined observations; $\nabla\Delta\rho^{i,j}_{\mathrm{a,b}}$ represents double-difference station-satellite distance; $m^{i,j}_{\mathrm{a,b}}$ represents tropospheric projection coefficient, $\Delta T_{\mathrm{a,b}}$ represents single-difference zenith tropospheric delay between stations; λ_{IF} represents ionospheric-free observation wavelength, $\nabla\Delta N^{i,j}_{\mathrm{a,b},IF}$ represents double-difference ionospheric-free combined ambiguity; $\varepsilon(\nabla\Delta P^{i,j}_{\mathrm{a,b},IF}), \varepsilon(\nabla\Delta L^{i,j}_{\mathrm{a,b},IF})$ represent pseudo-range and carrier observation noise, respectively.

Since the ionosphere-free (IF) ambiguity does not retain integer characteristics, it is typically split into wide-lane (WL) and narrow-lane (NL) components [15]. The WL ambiguity is resolved to an integer by averaging and rounding the Melbourne–Wübbena (MW) combination [16]. The corresponding fixing formula is given as follows:

$$\begin{aligned} \nabla\Delta N^{i,j}_{\mathrm{a,b,NL}} &= \frac{f_1 + f_2}{f_1}\nabla\Delta N^{i,j}_{\mathrm{a,b,IF}} - \frac{f_2}{f_1 - f_2}\nabla\Delta N^{i,j}_{\mathrm{a,b,WL}} \\ \nabla\Delta N^{i,j}_{\mathrm{a,b,WL}} &= \mathrm{int}<\nabla\Delta L^{i,j}_{\mathrm{a,b},MW}> \end{aligned} \tag{2}$$

Where, $\nabla\Delta N^{i,j}_{\mathrm{a,b,WL}}, \nabla\Delta N^{i,j}_{\mathrm{a,b,NL}}$ represents the double-difference wide-lane and narrow-lane ambiguity; f_1, f_2 represents the frequency of L1 and L2 observations respectively; int represents the rounding operation; $\nabla\Delta L^{i,j}_{\mathrm{a,b},MW}$ represents the double-difference MW combination observations; <·> represents smooth filtering between epochs. After the ambiguity is fixed, the tropospheric delay of double-difference $\nabla\Delta T^{i,j}_{\mathrm{a,b}}$ and the

Journal of Physics: Conference Series **3056** (2025) 012014
doi:10.1088/1742-6596/3056/1/012014

double-difference ionospheric delay $\nabla\Delta I_{\mathrm{a,b}}^{i,j}$ on L1 observations are extracted respectively, and the regional atmospheric delay model is established.

Considering the influence of height difference on the accuracy of tropospheric interpolation, the tropospheric interpolation model considering additional height correction (AHC) is adopted [17,18]. Firstly, by using the GPT2-1w model and GMF projection function [19], the baseline double- difference tropospheric delay correction is divided into distance dependent part and height dependent part, then $\nabla\Delta T_{\mathrm{a,b}}^{i,j}$ can be expressed as:

$$\nabla\Delta T_{\mathrm{a,b}}^{i,j} = \nabla\Delta T_{\mathrm{a,b}}^{i,j}(h_0) + dh(T_{a,b}) \tag{3}$$

$\nabla\Delta T_{\mathrm{a,b}}^{i,j}$ donates the baseline double-difference tropospheric delay, $\nabla\Delta T_{\mathrm{a,b}}^{i,j}(h_0)$ represents tropospheric delay in double-difference form of the reference height plane h_0, and $dh(T_{a,b})$ represents the height dependent part. According to the coordinates of the reference station $\vec{\mathrm{X}}_a = (x_a, \mathrm{y}_a, \mathrm{h}_a), \vec{\mathrm{X}}_b = (x_b, \mathrm{y}_b, \mathrm{h}_b), \vec{\mathrm{X}}_c = (x_c, \mathrm{y}_c, \mathrm{h}_c)$ and VRS $\vec{\mathrm{X}}_{\mathrm{vrs}} = (x_{vrs}, \mathrm{y}_{vrs}, \mathrm{h}_{vrs})$, using the linear interpolation model (LIM) and linear combination model (LCM) models [20], respectively, the distance dependent part $\nabla\Delta T_{\mathrm{a,vrs}}^{i,j}(h_0)$ on the reference height surface h_0 and the height dependent part $dh(T_{a,\mathrm{vrs}})$ of the VRS can be interpolated as follows:

$$\begin{aligned} &\nabla\Delta T_{\mathrm{a,vrs}}^{i,j}(h_0) = (x_{vrs} - x_a \quad y_{vrs} - y_a)\begin{pmatrix} x_b - x_a & y_b - y_a \\ x_c - x_a & y_c - y_a \end{pmatrix}^{-1} \begin{pmatrix} \nabla\Delta T_{\mathrm{a,b}}^{i,j}(h_0) \\ \nabla\Delta T_{\mathrm{a,c}}^{i,j}(h_0) \end{pmatrix} \\ &dh(T_{a,\mathrm{vrs}}) = \alpha_1 dh(T_{a,\mathrm{b}}) + \alpha_2 dh(T_{a,\mathrm{c}}) \end{aligned} \tag{4}$$

Where, the coefficient $\alpha_i (i = 1, 2, 3)$ satisfies the following constraints:

$$\sum_{i=1}^{3}\alpha_i = 1, \sum_{i=1}^{3}\alpha_i(\vec{\mathrm{X}}_{vrs} - \vec{\mathrm{X}}_i) = 0, \sum_{i=1}^{3}\alpha_i^2 = \min \tag{5}$$

Then the double-difference tropospheric delay of the VRS can be expressed as:

$$\nabla\Delta T_{\mathrm{a,vrs}}^{i,j} = \nabla\Delta T_{\mathrm{a,vrs}}^{i,j}(h_0) + dh(T_{a,\mathrm{vrs}}) \tag{6}$$

In ionospheric interpolation, the LIM model is employed to interpolate the double-difference ionospheric delay to obtain the double-difference ionospheric delay of the VRS [20].

$$\nabla\Delta I^{i,j}_{\mathrm{a,vrs}} = (x_{vrs} - x_a \quad y_{vrs} - y_a)\begin{pmatrix} x_b - x_a & y_b - y_a \\ x_c - x_a & y_c - y_a \end{pmatrix}^{-1}\begin{pmatrix} \nabla\Delta I^{i,j}_{\mathrm{a,b}} \\ \nabla\Delta I^{i,j}_{\mathrm{a,c}} \end{pmatrix} \tag{7}$$

If the VRS atmospheric delay is determined via double differencing, the observation of the VRS[21] is expressed as follows:

$$\begin{aligned} P^i_{v,k} &= P^i_{a,k} + \Delta\rho^i_{a,v} + (\nabla\Delta I^{i,\mathrm{ref}}_{a,v} + \Delta I^{ref}_{a,v}) + (\nabla\Delta T^{i,\mathrm{ref}}_{a,v} + \Delta T^{ref}_{a,v}) \\ \Phi^i_{v,k} &= \Phi^i_{a,k} + \frac{1}{\lambda_k}\Delta\rho^i_{a,v} - \frac{1}{\lambda_k}(\nabla\Delta I^{i,\mathrm{ref}}_{a,v} + \Delta I^{ref}_{a,v}) + \frac{1}{\lambda_k}(\nabla\Delta T^{i,\mathrm{ref}}_{a,v} + \Delta T^{ref}_{a,v}) \end{aligned} \tag{8}$$

2.2 Data processing on the monitoring-end

When the monitoring end data is processed, the baseline data process is first solved, after which VRS observables derived in the previous step are merged with the monitoring-receiver measurements to construct double-difference data. Although the virtual reference point is physically only tens to a few hundred metres from the monitored target, considering that the VRS observation is actually calculated by the CORS station which is tens of kilometers away from the monitoring point, and the vertical separation between the CORS and monitoring site can be on the order of several hundred meters, thus tropospheric and ionospheric residuals introduced by double differencing combinations can't be neglected. In this paper, the L1 and L2 dual-frequency un-difference models are used for the baseline data process and observation equation is expressed as:

$$\begin{aligned} \nabla\Delta P^{i,j}_{\mathrm{a,b,1}} &= \nabla\Delta\rho^{i,j}_{\mathrm{a,b}} + \nabla\Delta I^{i,j}_{\mathrm{a,b}} + m^{i,j}_{\mathrm{a,b}} \times \Delta T_{\mathrm{a,b}} + \varepsilon(\nabla\Delta P^{i,j}_{\mathrm{a,b,1}}) \\ \nabla\Delta P^{i,j}_{\mathrm{a,b,2}} &= \nabla\Delta\rho^{i,j}_{\mathrm{a,b}} + \mu \times \nabla\Delta I^{i,j}_{\mathrm{a,b}} + m^{i,j}_{\mathrm{a,b}} \times \Delta T_{\mathrm{a,b}} + \varepsilon(\nabla\Delta P^{i,j}_{\mathrm{a,b,2}}) \\ \nabla\Delta L^{i,j}_{\mathrm{a,b,1}} &= \nabla\Delta\rho^{i,j}_{\mathrm{a,b}} - \nabla\Delta I^{i,j}_{\mathrm{a,b}} + m^{i,j}_{\mathrm{a,b}} \times \Delta T_{\mathrm{a,b}} + \lambda_1\nabla\Delta N^{i,j}_{\mathrm{a,b,1}} + \varepsilon(\nabla\Delta L^{i,j}_{\mathrm{a,b,1}}) \\ \nabla\Delta L^{i,j}_{\mathrm{a,b,2}} &= \nabla\Delta\rho^{i,j}_{\mathrm{a,b}} - \mu \times \nabla\Delta I^{i,j}_{\mathrm{a,b}} + m^{i,j}_{\mathrm{a,b}} \times \Delta T_{\mathrm{a,b}} + \lambda_2\nabla\Delta N^{i,j}_{\mathrm{a,b,2}} + \varepsilon(\nabla\Delta L^{i,j}_{\mathrm{a,b,2}}) \end{aligned} \tag{9}$$

Where, $\nabla\Delta P^{i,j}_{\mathrm{a,b,1}}, \nabla\Delta P^{i,j}_{\mathrm{a,b,2}}$ represent L1 and L2 double-difference pseudo-range observations respectively, $\nabla\Delta L^{i,j}_{\mathrm{a,b,1}}, \nabla\Delta L^{i,j}_{\mathrm{a,b,2}}$ denote the double-differenced carrier-phase measurements on L1 and L2 frequencies, respectively, μ represents the ionospheric proportional coefficient of L2 relative to L1 observations, $\nabla\Delta I^{i,j}_{\mathrm{a,b}}$ denotes the ionospheric delay in double-differenced form at the L1 frequency observations, λ_1, λ_2 represents the wave lengths of L1 and L2 observations, $\nabla\Delta N^{i,j}_{\mathrm{a,b,1}}, \nabla\Delta N^{i,j}_{\mathrm{a,b,2}}$ represents the double-difference ambiguities of L1 and L2 observations, respectively, and $\varepsilon(\nabla\Delta P^{i,j}_{\mathrm{a,b,1}}), \varepsilon(\nabla\Delta P^{i,j}_{\mathrm{a,b,2}})$ and $\varepsilon(\nabla\Delta L^{i,j}_{\mathrm{a,b,1}}), \varepsilon(\nabla\Delta L^{i,j}_{\mathrm{a,b,2}})$ represent the noise of observations.

In order to facilitate data processing, Kalman filtering is used to solve the problem. In Eq. (9), the parameters to be estimated include baseline vector, single difference zenith tropospheric delay, double-difference slant ionospheric delay and double-difference ambiguity

Journal of Physics: Conference Series **3056** (2025) 012014 doi:10.1088/1742-6596/3056/1/012014

parameters. The baseline vector is considered to remain unchanged in one solution. The parameters of tropospheric delay and double-difference ionospheric are considered as a random walk, and the probability spectral densities are $2cm/\sqrt{h}$ and $30cm/\sqrt{h}$, respectively[22].

After the baseline is solved, to further improve the accuracy of data process, the VRS and the monitoring stations are then combined to form a delaunay triangulation network and the triangulation constraint is added for network adjustment. This constraint is mathematically represented by:

$$\nabla\Delta X_{a,b} + \nabla\Delta X_{VRS,a} + \nabla\Delta X_{b,VRS} = 0 \tag{10}$$

Where $\nabla\Delta X_{a,b}$, $\nabla\Delta X_{VRS,a}$ and $\nabla\Delta X_{b,VRS}$ represent the double-difference baseline vectors of VRS and monitoring stations, respectively.

3. Experimental verification and analysis

This study conducts experimental area is in the south of China, where the mountains are continuous and the terrain is extremely complex, with the highest altitude of 2590 meters and the lowest altitude of 154 meters. Affected by the warm and humid air flow of the southeast ocean, the territory is humid and rainy, and there are many hidden danger points of geological disasters. A total of 90 monitoring stations in 30 hidden danger points of landslides are selected for experimental analysis. 6 individual CORS stations have been built in and around the experimental area, with an average station spacing of about 30 km and the information and distribution of CORS stations are illustrated in list 1 and Fig. 1.

Table 1. CORS station information list

Station Name	Longitude(E/degree)	Latitude(N/degree)	Height(m)
JZ01	101.531003850	26.441767940	1959.745
JZ02	101.821517997	26.256150600	1665.148
JZ03	101.490584972	26.166092211	1399.214
JZ04	101.772626667	26.048576741	636.061
JZ05	101.597984535	26.000589335	1583.311
JZ06	101.939514045	25.981663991	1030.966

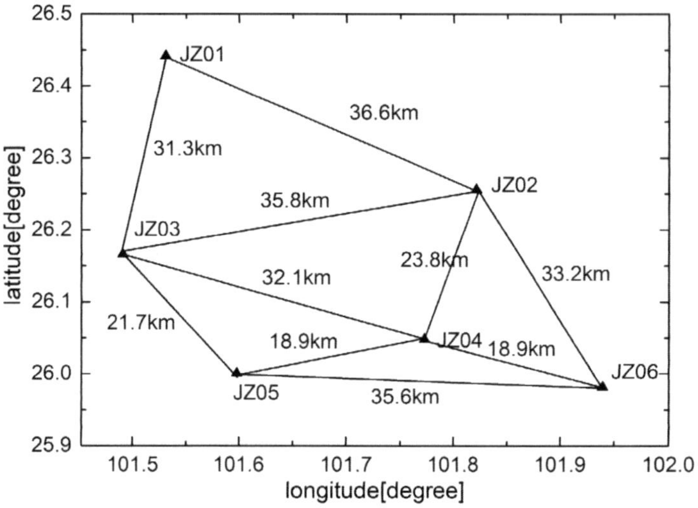

Figure 1. Distribution of the CORS stations

3.1. Analysis of the server-end data process

The quality of VRS positioning is largely governed by how well ionospheric and tropospheric delays are interpolated; therefore, this study evaluates the performance of several delay-interpolation models. Firstly, a subset of monitoring-stastion coordinates are selected, and both the LIM and AHC-LIM models are employed to interpolate double-differenced ionospheric and tropospheric delays between the VRS and the main reference site. Simultaneously, observation data from the monitoring receiver are used to compute the double-differenced ionospheric and tropospheric delays for the corresponding satellite. The resulting discrepancy is then analyzed to evaluate the accuracy of the ionospheric and tropospheric delay models.

The ionospheric delay exhibits significant diurnal variation—most prominent at midday and relatively stable at night. Figure 2 compares the performance of double-difference ionospheric interpolation during daytime and nighttime periods. During nighttime, interpolated double-difference ionospheric delays closely match those computed from actual measurements. In contrast, strong ionospheric activity at noon reduces interpolation accuracy, especially at low satellite elevation angles, where errors may exceed half an L1 wavelength cycle.

Figure 3 presents the statistical deviations between interpolated and computed double-differenced ionospheric delays across all satellites during both midday and nighttime periods. As illustrated in Figure 3, nighttime ionospheric delay interpolation yields a mean error of just 0.6 cm, with a STD of 2.7 cm. The double-difference interpolation performs well in mitigating ionospheric effects during nighttime. At noon, the average interpolation error rises to 1.3 cm, with a notably high STD of 5.4 cm. In extreme cases, errors may exceed 20 cm, adversely impacting ambiguity resolution at the monitoring site.

Hence, at the monitoring station, the double-difference ionospheric delay must be either modeled as a parameter or mitigated using dual-frequency ionosphere-free combinations.

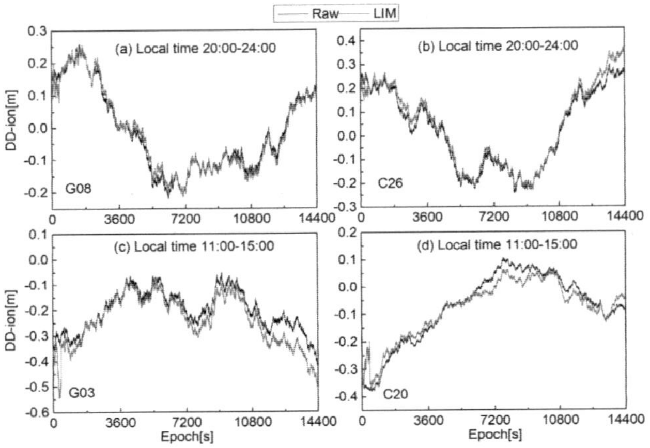

Figure 2.Temporal comparison of double-differenced ionospheric model accuracy

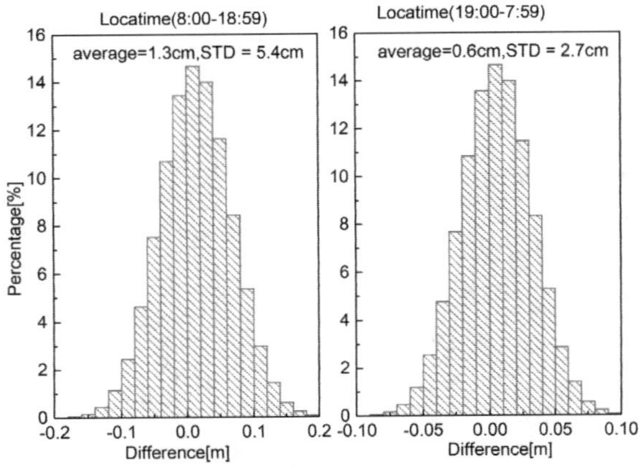

Figure 3.Statistical evaluation of double-difference ionospheric modeling across different time intervals

Tropospheric delay exhibits a strong dependence on station elevation. Therefore, interpolation accuracy under varying inter-station elevation differences is analyzed for the double-difference tropospheric delay. Figure 4 illustrates the interpolation performance of the LIM and AHC-LIM models for station height differences of 100 m, 400 m, 600 m, and 800 m, respectively.

As indicated in Fig. 4, for elevation differences below 100 m, both LIM and AHC-LIM models exhibit similar interpolation accuracy. However, with increasing elevation disparity, the LIM model introduces noticeable systematic deviations in tropospheric delay interpolation. In contrast, the AHC-LIM model maintains consistency with actual double-differenced delay values and does not exhibit evident systematic bias.

Figure 5 presents the overall interpolation accuracy across all satellites for both LIM and AHC-LIM models under varying height conditions. As demonstrated in Fig. 5, the AHC-LIM model significantly enhances tropospheric delay interpolation accuracy, but for large station elevation gaps, the mean interpolation error still reaches several centimeters, posing challenges for achieving millimeter-level precision required in landslide monitoring. Hence, in regions with substantial height disparities, residual tropospheric delay errors must be explicitly modeled as parameters during monitoring-end processing.

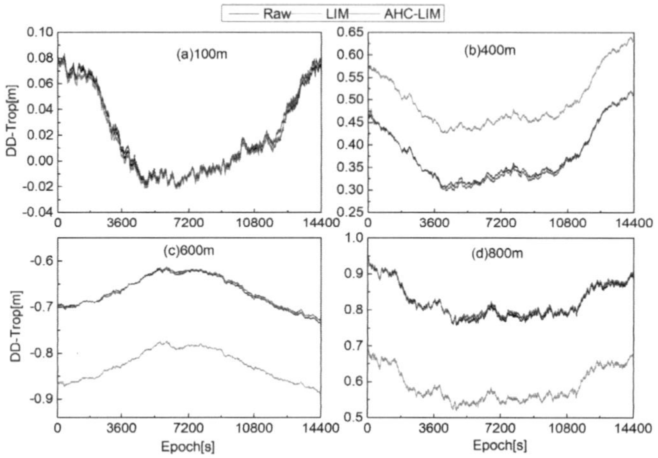

Figure 4.Accuracy of the LIM and the AHC-LIM model with different height differences

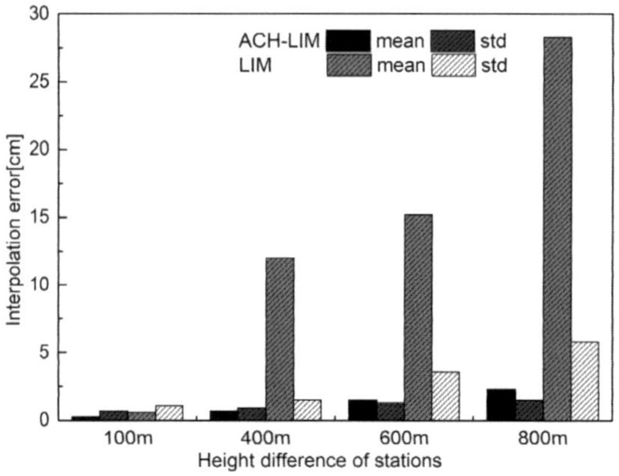

Figure 5.Accuracy statistics of all satellites with different height differences of the LIM and AHC-LIM model

3.3 Analysis of the monitoring-end data process

Observation data from each monitoring station are processed using both the proposed method and a conventional approach. Initially, the ambiguity resolution success rates for both methods are evaluated. Table 2 summarizes the comparative statistics of ambiguity resolution performance for the two approaches, where the ambiguity resolution rate is defined as the proportion of epochs with correctly fixed ambiguities to the total number of satellite epochs.

Table 2.Statistical comparison of ambiguity resolution success rates between the two methods

Methods	Ambiguity Fixing Success Rate[%]		
	Min. Value	Max. Value	Mean
traditional method	98.5	99.99	99.3

proposed method	97.3	99.5	98.6

As shown in Table 2, the traditional method effectively suppresses various error sources due to its short baseline and accurate initial station coordinates. As a result, its ambiguity resolution rate is high, with minimum and mean values reaching 98.5% and 99.3%, respectively.

While the proposed method yields slightly lower performance, it still achieves a minimum of 97.3% and an average of 98.6%, demonstrating that VRS-derived data remain sufficient for reliable double-difference ambiguity resolution.

Figure 6 illustrates the statistical distribution of double-difference residuals for both the traditional and proposed methods. As depicted in Fig. 6, residuals from both methods follow a normal distribution. The standard deviations of phase and pseudo-range residuals were 0.005 m and 0.564 m for the traditional method, and 0.009 m and 0.847 m for the proposed method. This indicates that the proposed method is also capable of mitigating baseline-related errors to a considerable extent.

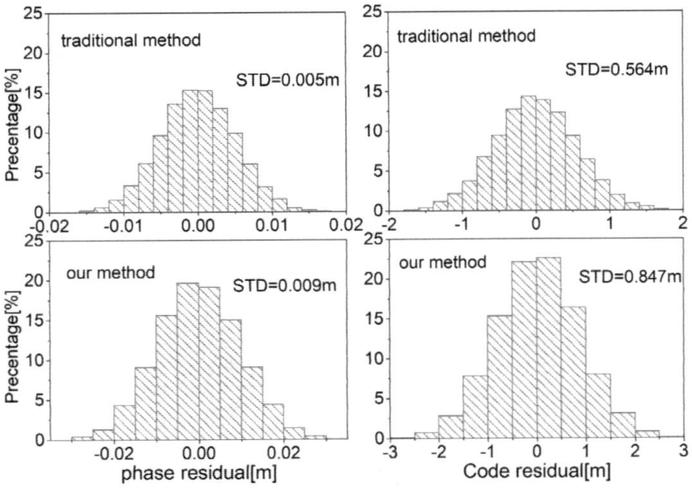

Figure 6.Statistics of residual distribution of double-difference pseudorange and phase observations

Observation data collected over one week from 60 monitoring stations are segmented into groups with varying durations: 2, 4, 6, 8, 10, and 12 hours. Baseline processing is performed using the proposed method, and root mean square (RMS) positioning errors are computed in the N, E, and U directions. The corresponding results are presented in Figure 7.

The positioning accuracy is strongly influenced by the length of the observation period. With only 2 hours of data, accuracy is limited to the centimeter level, and the average vertical RMS reaches approximately 13 mm. As observation time extends, positioning precision improves progressively, reaching sub-5 mm horizontal and sub-10 mm vertical RMS once the duration exceeds 4 hours.

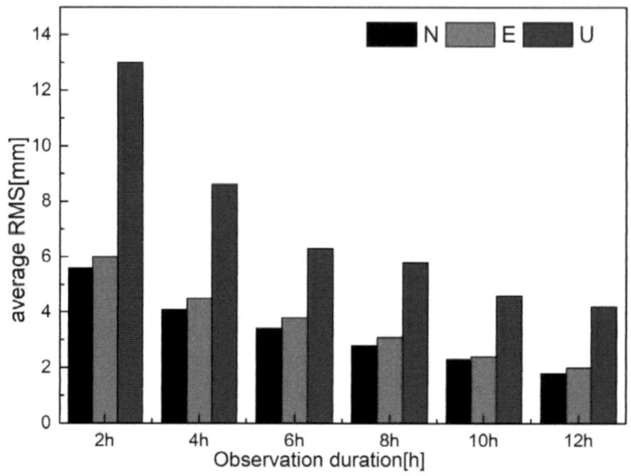

Figure 7.The changes in accuracy of the results of data with different observation durations

To optimize the accuracy of the processing workflow, an additional correction was applied, the 4h baseline solution results of the proposed method are processed by network adjustment. Figure 8 shows the RMS statistics for all baselines before and after the net adjustment. As shown in Fig.8, after the network adjustment, the average RMS of the baseline solution in the north, east, and up directions decrease from 4.9 mm, 5.5 mm and 9.8 mm to 4.6 mm, 5.1 mm and 9.2 mm, which are improved by 6%, 7% and 6% respectively.

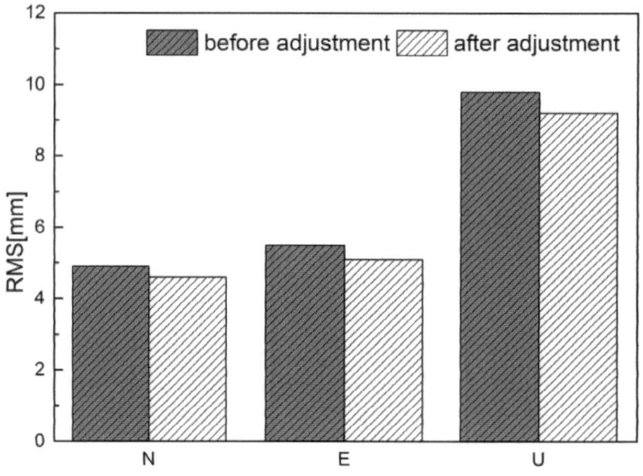

Figure 8.RMS statistics of all baselines before and after network adjustment

Figure 9 illustrates 15-day time series observations at a monitoring site using both the proposed and traditional methods. As indicated in Figure 9, the traditional method yields higher positioning accuracy—attributed to the short baseline—with RMS values of 3.4 mm, 4.3 mm, and 6.1 mm in the north, east, and vertical directions, respectively. Although slightly less accurate, the proposed method achieves RMS values of 5.1 mm, 6.0 mm, and 8.4 mm in the respective directions, which remains sufficient for millimeter-level deformation monitoring tasks.

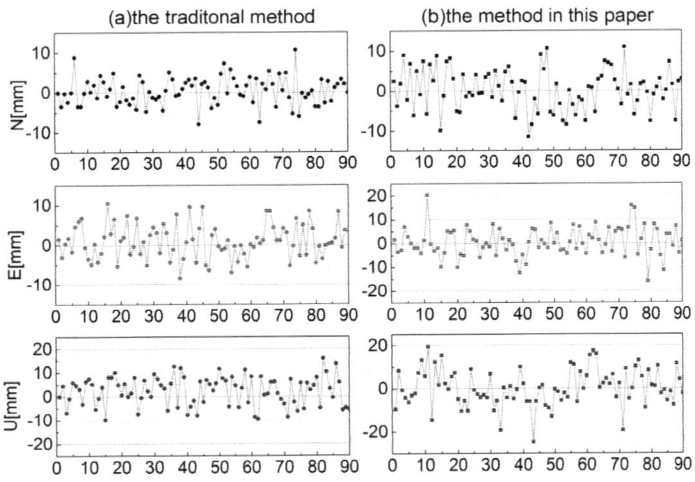

Figure 9.Time series of positioning results at the monitoring station

4. Conclusion

This study proposes an enhanced GNSS deformation-monitoring approach utilizing VRS, constructed directly from existing CORS measurements, eliminating the need for physical reference stations while maintaining high-precision positioning. The experimental analysis is carried out on about 30 landslide monitoring points in the south of China. Results indicate that:1) the proposed method achieves horizontal accuracies within 5 mm and vertical accuracies within 10 mm, provided observation sessions exceed four hours, thus fulfilling requirements for landslide monitoring and similar precision applications; 2) in low-latitude areas, strong midday ionospheric disturbances significantly impact interpolation accuracy; thus, double-difference ionospheric errors must be carefully addressed during monitoring-end data processing; and 3) When significant height differences exist between the monitoring site and the CORS reference stations, double-difference tropospheric delays become non-negligible and should be explicitly estimated as model parameters at the monitoring end. The proposed method uses several established CORS instead of building new monitoring reference stations, which can greatly reduce the application cost of deformation monitoring, and the VRS are generated by using multiple reference stations, which can overcome the problem of the changes of monitoring datum caused by the instability of a single reference station.

Acknowledgements

This work was supported in part by National Key R&D Program of China (2023YFB3907300); in part by The National Natural Science Foundation of China (U23A20280); and in part by Major Program (JD) of Hubei Province (2023BAA026).

Reference

[1] Zumberge J F, Heflin M B, Jefferson D C, et al. Precise Point Positioning for the Efficient and Robust Analysis of GPS Data from Large Networks[J]. Journal of Geophysical Research: Solid Earth, 1997, 102(B3): 5005-5017.

[2] Bisnath, S. and Y. Gao. Current State of Precise Point Positioning and Future Prospects and

Limitations. in: Sideris, M.G. (Ed.), Observing Our Changing Earth. Springer-Verlag, New York, pp. 615–623, 2008.

[3] Gili J A, Corominas J, Rius J. Using Global Positioning System Techniques in Landslide Monitoring[J]. Engineering Geology, 2000, 55(3): 167-192

[4] Rawat M S, Joshi V, Rawat B S, et al. Landslide Movement Monitoring Using GPS Technology: A Case Study of Bakthang Landslide, Gangtok, East Sikkim, India[J]. Journal of Development and Agricultural Economics, 2011, 3: 194-200

[5] Squarzoni C, Delacourt C, Allemand P. Differential single-frequency GPS monitoring of the LaValette landslide (French Alps)[J]. Engineering Geology, 2005, 79(3-4):215-229.

[6] Cheng, L.K. Liu, S.Y. Guan, B.O et al. Dynamic load monitoring of the Tsing Ma Bridge using a high-speed FBG sensor system. In Proceedings 2nd European Workshop on Structural Health Monitoring, Munich, Germany (2004)

[7] Rawat M S, Joshi V, Rawat B S, et al. Landslide Movement Monitoring Using GPS Technology: A Case Study of Bakthang Landslide, Gangtok, East Sikkim, India[J]. Journal of Development and Agricultural Economics, 2011, 3: 194-200.

[8] Kaloop M R, Wan H J. Dynamic Performance Analysis of the Towers of a Long-Span Bridge Based on GPS Monitoring Technique[J]. Journal of Sensors, 2016, 2016(4):1-14.

[9] D'Oreye N, Arjona A, Bagalwa M, et al. A Permanent Geodetic GNSS Network to Monitor Ground Deformation in the Virunga Volcanic Province[C]// Active Volcanism & Continental Rifting (AVCOR), Gisenyi, Rwanda, 2013

[10] Yusup A, Othman R, Musliman I A, et al. ISKANDARnet CORS Network Integrity Monitoring[J]. Jurnal Teknologi, 2014, 71(4): 11-19

[11] Uzel T, Eren K, Gulal E, et al. Monitoring the Tectonic Plate Movements in Turkey Based on the National Continuous GNSS Network[J]. Arabian Journal of Geosciences, 2013, 6(9): 3573-3580

[12] Zhang Mingzhi, Zhan Bing, Zhao Wenyi, et al. Design and Practice of High Precision Landslide Displacement Monitoring System Based on VRS[J]. The Chinese Journal of Geological Hazard and Control, 2020, 31(6): 54-59.

[13] Yuan Y B, Mi X L, Zhang B C. Initial Assessment of Single- and Dual-Frequency BDS-3 RTK Positioning[J]. Satellite Navigation, 2020, 1: 31

[14] Kozlov D, Tkachenko M. Centimeter-Level, Real-Time Kinematic Positioning with GPS+GLONASS C/a Receivers[J]. Navigation, 1998, 45(2): 137-147.

[15] Ge M, Gendt G, Rothacher M, et al. Resolution of GPS Carrier-Phase Ambiguities in Precise Point Positioning (PPP) with Daily Observations[J]. Journal of Geodesy, 2008, 82(7): 389-399

[16] Geng J H, Shi C, Ge M R, et al. Improving the Estimation of Fractional-Cycle Biases for Ambiguity Resolution in Precise Point Positioning[J]. Journal of Geodesy, 2012, 86(8): 579-589.

[17] Song J, Kee C, Park B, et al. Correction combination of compact network RTK considering tropospheric delay variation over height[C]//2014 IEEE/ION Position, Location and Navigation Symposium-PLANS 2014. IEEE, 2014: 95-101.

[18] Song J, Park B, Kee C. Comparative analysis of height-related multiple correction interpolation methods with constraints for network RTK in mountainous areas[J]. The Journal of Navigation, 2016, 69(5): 991-1010.

[19] Guo Q, Wu X. A global assessment of ray-traced and blind tropospheric models in the retrieval of tropospheric parameters from ground-based GPS observations[J]. Journal of Atmospheric and Solar-Terrestrial Physics, 2019, 190: 16-35.

[20] Song Weiwei, He Chengpeng, Gu Shengfeng. Performance Analysis of Ionospheric Enhanced PPP-RTK in Different Latitudes[J]. Geomatics and Information Science of Wuhan University, 2021, 46(12): 1832-1842.

[21] Wei E, Chai H, An Z, et al. VRS virtual observations generation algorithm[J]. Journal of Global Positioning Systems, 2006, 5(1-2): 76-81.

[22] Guo H, He H, Li J, et al. Estimation and mitigation of the main errors for centimetre-level compass RTK solutions over medium-long baselines[J]. The Journal of Navigation, 2011, 64(S1): S113-S126.

Passive Strain Sensor Based on RFID Patch Antenna

Shuo Yang[1], Xiaochen Wang[1*], Qiaozhi Song[2], Xiaochuan Liu[2], Peng Li[1], Yiren Yang[1]

[1] Southwest Jiaotong University,Chengdu,Sichuan Province, 610000, PR China

[2] National Key Laboratory of Strength and Structural Integrity, China Aircraft Strength Research Institute, Xi'an 710065, China

*Corresponding author's e-mail: wangxiaochen@swjtu.edu.cn

Abstract. With the rapid development of engineering construction, the demand for structural health monitoring is increasing day by day. Traditional structural monitoring techniques are confronted with problems such as cumbersome operation, high cost and low accuracy in practical applications. Structural health monitoring based on RFID technology provides a new solution. Through theoretical analysis and simulation in this paper, a passive wireless strain sensor based on RFID technology is designed, revealing the linear relationship between the resonant frequency offset of the antenna strain sensor and the strain, laying a theoretical foundation for subsequent experiments. Further, by optimizing the structure of the tag antenna, the low-cost manufacturing of the RFID strain sensor was successfully achieved, providing theoretical support for its large-scale application.

1. Introduction

In recent years, the rapid advancement of engineering disciplines such as civil infrastructure, aerospace, and architectural systems has propelled structural health monitoring (SHM) into a cornerstone technology for ensuring the safety and durability of critical infrastructure. Conventional monitoring techniques—including resistive strain gauges and fiber-optic sensors—are constrained by their reliance on wired connectivity and external power supplies, leading to challenges such as complex cabling architectures, prohibitive maintenance costs, and insufficient long-term reliability[1]-[3]. These limitations are particularly pronounced in large-scale, distributed monitoring scenarios, where the logistical and financial burdens of wired systems become untenable. Consequently, the development of intelligent sensing technologies characterized by low cost, passive operation, and wireless data transmission has emerged as a pivotal innovation pathway, driving transformative advancements in structural monitoring frameworks to address scalability, flexibility, and sustainability demands.

Radio Frequency Identification (RFID) technology, leveraging its inherent advantages of passive operation, environmental resilience, and mass deployment potential, offers a transformative pathway for structural health monitoring (SHM). While conventional RFID tags are primarily utilized for identity recognition and data storage, the resonant frequency shift of their antenna structures under mechanical deformation provides a physical foundation for strain sensing. Recent years have witnessed burgeoning research in RFID antenna-based strain sensing, where the coupling between antenna deformation and electromagnetic properties enables passive, wireless strain detection[4]-[8].

The U Tata group pioneered theoretical and experimental investigations into the resonant frequency-strain relationship of rectangular microstrip patch antennas[4]. Since then, research on RFID-based strain sensors has proliferated. Prof. Occhiuzzi's team at the University of Rome, Italy, employed dipole antennas for strain detection, utilizing antenna gain as the monitoring parameter[5]. In contrast, Zhang et al. developed a flexible RFID sensor for simultaneous strain and humidity monitoring, employing a backpropagation (BP) neural network to decouple their coupled signal responses[9]. Xie et al. designed a dual-axis (transverse and longitudinal) wireless RFID strain sensor; however, its reliance on numerous shorting vias incurred elevated manufacturing costs, hindering scalability for large-scale engineering applications[10].

Building upon the aforementioned studies, this work focuses on developing a passive strain sensor based on RFID patch antennas, proposing a strain transduction mechanism that directly correlates patch deformation with resonant frequency shifts. Through rigorous analysis grounded in electromagnetic theory and strain modeling of microstrip antennas, we elucidate the linear relationship between the resonant frequency shift and applied strain, establishing a quantitative expression relating sensitivity to structural parameters. Leveraging this framework, multiphysics-coupled modeling of the tag antenna was implemented using high-frequency electromagnetic simulation software (HFSS) to optimize geometric parameters such as radiation patch dimensions, substrate materials, and feed configurations. This optimization achieves significant reductions in manufacturing complexity while retaining high strain sensitivity, thereby advancing practical deployability for large-scale structural monitoring.

2. Working principle of RFID strain sensors

As a core automated identification technology in modern IoT perception systems, Radio Frequency Identification (RFID) operates on electromagnetic wave energy transfer principles. It establishes bidirectional communication links through inductive near-field coupling or electromagnetic backscattering techniques within designated frequency bands, enabling remote target authentication and motion trajectory tracking. In the ultra-high frequency (UHF, 860 – 960 MHz) band, RFID systems leverage backscattering mechanisms to construct information transmission channels: interrogator-emitted radio signals are reflected and modulated by tag antennas, generating carrier-modulated responses that facilitate passive, contactless data exchange. Driven by advancements in semiconductor fabrication and economies of scale, RFID technology has evolved over decades into a standardized protocol architecture. Its passive tag design simplifies hardware structures while significantly reducing deployment and maintenance costs, thereby supporting scalable distributed sensor networks through rapid deployment and flexible configuration. This technological maturation underscores RFID's pivotal role in enabling cost-effective, large-scale IoT infrastructure with robust sensing and identification capabilities.

An RFID system comprises three core components: an RFID tag, an interrogator (reader), and a data processing subsystem. In structural health monitoring (SHM) applications, the RFID tag integrates a sensing target with a tag antenna. Within the SHM framework, the sensing node serves as a critical transducer, converting mechanical responses of monitored structures—such as strain fields and thermal gradients—into quantifiable optical, electrical, or acoustic signals via electromechanical coupling effects. Building on this mechanism, the RFID tag antenna is reconfigured as a passive sensing element: external loads inducing substrate deformation or temperature variations altering material permittivity trigger measurable deviations in electromagnetic response parameters, including resonant frequency shifts, impedance spectrum variations, and backscattering gain deviations. These time-variant parameters are encoded with the tag's unique identifier through modulation-demodulation processes, forming composite data packets. Following quadrature down-conversion in the RF front-end, structural parameters are wirelessly and passively sensed through signal demodulation and feature extraction algorithms. In SHM applications, this approach enables the RFID tag antenna to detect deformation- or temperature-induced electrical signal alterations (e.g., frequency, phase) and transmit them wirelessly, integrated with its embedded identification code, for comprehensive structural state assessment.

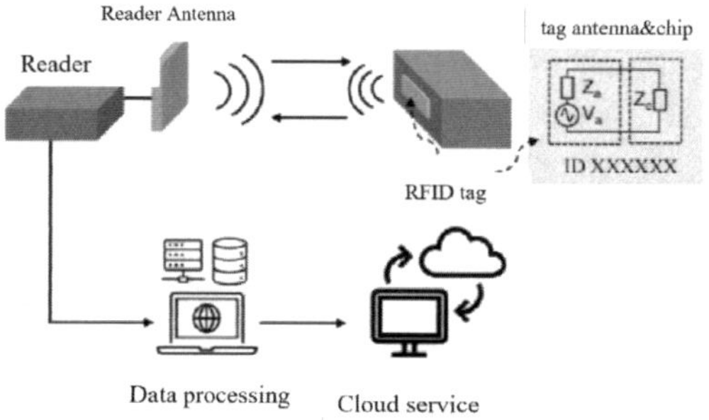

Figure 1. RFID System Composition[11]

2.1. Strain measurement principle of RFID antenna

The rectangular microstrip patch antenna was selected for its grounded configuration, which demonstrates superior efficacy in metallic strain monitoring applications. Based on this rationale, the proposed antenna in this study adopts a rectangular microstrip topology. According to the transmission line model, the initial resonant frequency of the rectangular antenna can be expressed as:

$$f_r = \frac{c}{2L_e\sqrt{\varepsilon_e}} = \frac{c}{2\sqrt{\varepsilon_e}\,(L+2\Delta l)} \tag{1}$$

In the equation, f_r represents the initial resonant frequency of the antenna; c corresponds to the speed of light in a vacuum; L_e denotes the effective electrical length; L is the effective dielectric constant; L signifies the physical antenna length; and Δl refers to the compensation length. Here, the effective dielectric constant and compensation length are dependent on the antenna's structural parameters. Their relationship is defined by

$$\Delta l = 0.412h\frac{(\varepsilon_e+0.3)|\frac{W}{h}+0.264|}{(\varepsilon_e-0.258)|\frac{W}{h}+0.813|} \tag{2}$$

$$\varepsilon_e = \frac{\varepsilon_r+1}{2}+\frac{\varepsilon_r-1}{2}|1+10\frac{h}{W}|^{-1/2} \tag{3}$$

In the governing equations, ε_e denotes the relative permittivity of the dielectric substrate, W represents the width of the antenna, and h corresponds to the thickness of the substrate. These parameters collectively define the electromagnetic characteristics and geometric configuration of the microstrip structure within the proposed design framework.

When the microstrip antenna is subjected to axial strain ε, the variation in the equivalent dielectric constant ε_e remains within a negligible range, whereas the antenna's geometric dimensions undergo deformation. Due to the Poisson effect, for an isotropic dielectric substrate, the length L, width W, and thickness h of the antenna are modulated as follows

$$L' = L(1+\varepsilon) \tag{4}$$

$$W' = W(1-\mu\varepsilon) \tag{5}$$

$$h' = h(1-\mu\varepsilon) \tag{6}$$

As derived from the preceding equation, the variations in equivalent dielectric constant ε_e and compensation length Δl exhibit negligible magnitudes and can thus be disregarded. Under these conditions, the resonant frequency of the antenna is expressed as:

$$f_r' = \frac{c}{2\sqrt{\varepsilon_e'}\,[(1+\varepsilon)L+2\Delta l]} \tag{7}$$

From the equation above, it can be derived that

$$\Delta f = -|\frac{f_r}{1+\varepsilon}|\varepsilon \approx -f_r\varepsilon = K\varepsilon \quad (8)$$

Here, K serves as the antenna sensitivity. As evident from the equation above, when the antenna undergoes axial tensile strain, its resonant frequency shifts toward lower frequencies; this frequency shift exhibits a linear relationship with the applied strain.

2.2. Impedance matching theory of RFID antenna

In RFID systems, the impedance matching between the microstrip antenna and the tag chip critically determines energy transfer efficiency. As illustrated by the equivalent circuit in the Figure 2, the antenna's equivalent source impedance under electromagnetic wave excitation is denoted as $Z_a = R_a + jX_a$, while the chip impedance is labeled as $Z_c = R_c + jX_c$. According to circuit theory, maximum power transfer is achieved when the source and load impedances are complex conjugates, a principle termed the conjugate matching or maximum power transfer theory. Defining the power transmission coefficient as τ, the relationship can be expressed as:

$$P_c = \tau P_a \quad (9)$$

In the equation, P_c and P_a represent the power delivered to the tag chip from the microstrip antenna and the power received by the tag antenna, respectively. When the tag antenna and chip achieve conjugate matching, the power transmission coefficient τ is maximized and can be formulated as

$$\tau = \frac{4R_aR_c}{|Z_a + Z_c|^2} = 1 - \Gamma^2 \quad (10)$$

In the formulation, Γ denotes the reflection coefficient, defined as the normalized power transmission coefficient, while τ characnterizes the energy reflection ratio. The return loss Γ^2 is related to the reflection coefficient S_{11} by the expression

$$S_{11} = 20\log|\Gamma| \quad (11)$$

As derived above, a smaller return loss S_{11} corresponds to higher energy transfer efficiency from the antenna to the chip and reduced reflected energy. At antenna resonance, the transmitted power reaches its maximum while the return loss S_{11} attains its minimum value. Consequently, the frequency corresponding to the minimum return loss is defined as the resonant frequency. In engineering applications, a return loss S_{11} below 10 dB—equivalent to a system energy loss of less than 10%—is deemed acceptable to meet operational specifications.

2.3. Detection principle of the resonant frequency of RFID antenna

When the reader transmits electromagnetic waves into free space via its antenna, and all conditions (free space, port, and polarization matching) are satisfied, this process can be modeled by the Friss transmission equation

$$P_t = \frac{P_r G_t}{4\pi R^2} * \frac{\lambda^2}{4\pi} G_r \quad (12)$$

In the equation, P_t and P_r denote the power received by the antenna and the power transmitted by the reader, respectively. G_r and G_t represent the antenna gains of the reader and the tag, while R corresponds to the distance between the reader and tag antennas.

Journal of Physics: Conference Series **3056** (2025) 012015 doi:10.1088/1742-6596/3056/1/012015

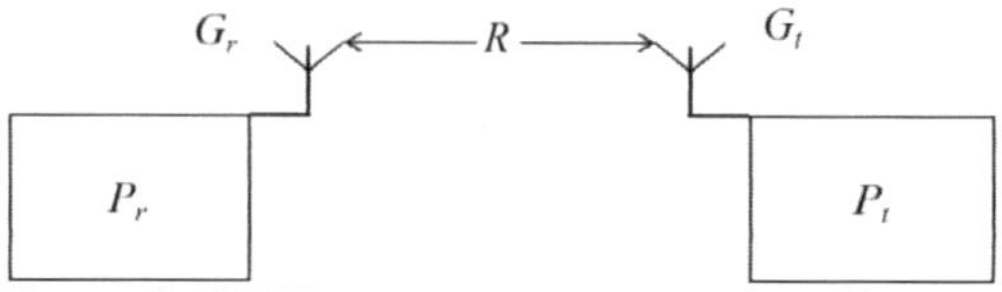

Figure 2. RFID tag antenna reception diagram

When the power received by the tag antenna is propagated to the tag chip, a portion is reflected back, while the remaining portion is harvested by the chip. The magnitude of the received power can be expressed as

$$P_c = \tau P_t \tag{13}$$

Due to the presence of capacitive components and other circuitry within the tag chip, a minimum power requirement (denoted as the threshold activation power P_{th}) is necessary to activate its operation. Therefore, the relationship between the minimum received power at the tag antenna P_t and the threshold activation power of the chip P_{th} is defined as

$$P_{th} = \tau P_t \tag{14}$$

As derived from the equation above, the reader transmit power P_r and the tag threshold activation power P_{th} must satisfy the following relationship

$$P_r = \frac{(4\pi R)^2 P_{th}}{\lambda^2 G_r G_t \left(1-|\Gamma|^2\right)} \tag{15}$$

As demonstrated by the equation above, the resonant power of the antenna can be indirectly ascertained by monitoring the threshold power of the chip.

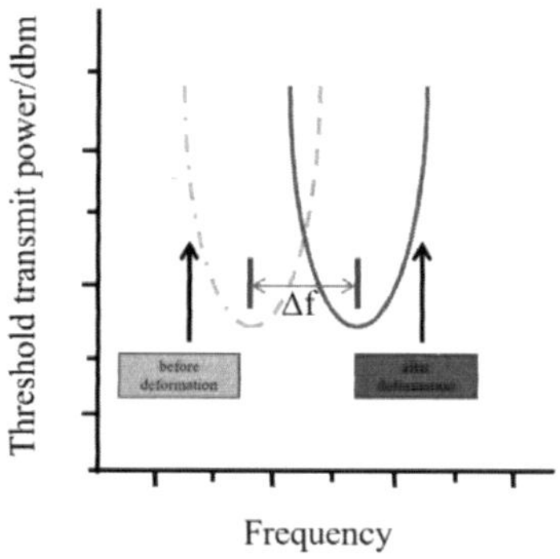

Figure 3. The relationship between threshold transmission power and transmission frequency

3. Simulation of RFID tag antenna

3.1 Antenna design and optimization

Based on the theoretical formulation for rectangular microstrip patch antennas, the substrate material is selected as FR4 (glass-reinforced epoxy laminate), which exhibits favorable electrical properties and mechanical robustness, making it widely adopted in RF applications. The substrate's relative permittivity is 4.4, with a Poisson's ratio of 0.15. The Higgs-3 chip is employed, presenting an impedance of 26.9-199*j* Ohm at 925 MHz, as illustrated in the Figure 4. The designed antenna model and its parameters are depicted in the Figure 5, with the impedance matching mechanism utilizing an embedded slot matching technique. Unlike conventional short-circuited stub matching, this approach simplifies fabrication processes for tag antennas, reducing manufacturing complexity. By modulating the Gap dimension, precise impedance matching is achieved, endowing this methodology with superior impedance adaptability.

Operating Conditions & Electrical Characteristics

Symbol	Parameter	Conditions / Capability	Min	Typ	Max	Units
Operating Conditions						
T_A	Operating Temperature		-50		+85	°C
f_{op}	Operating Frequency		860		960	MHz
Electrical Characteristics						
S_R	Sensitivity during Read	With 2dBi directivity of dipole		-20		dBm
I_S	Interference Signal Suppression			-4		dB
R_P	Equivalent input parallel resistance	At -14 dBm input power		1500		Ohms
C_P	Equivalent input parallel Capacitance	At -14 dBm input power		0.85		pF
D_{ret}	Data Retention		50			Years
P_{cycl}	Programming Cycles at 25°C			100,000		Cycles

Figure 4. Higgs-3 Data diagram

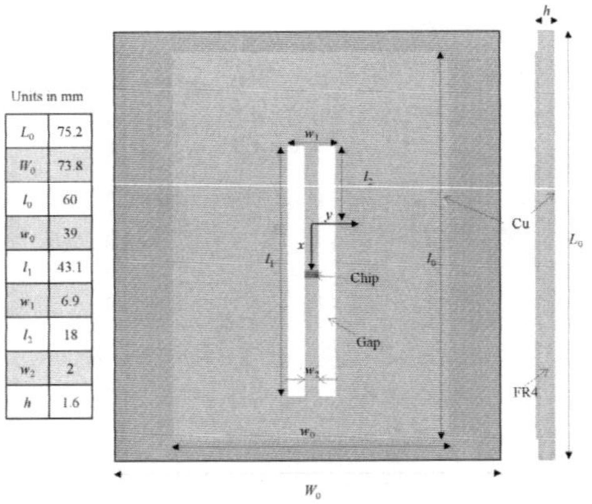

Figure 5. Initial antenna configuration and dimension

However, the initial patch area exceeds practical constraints for RFID sensors, where reduced patch area correlates with enhanced sensing accuracy. To address this, five slots are introduced on both width edges of the antenna, as illustrated in the Figure 6. By integrating serrated slots, the surface current path of the patch antenna is strategically prolonged, enabling a 48% reduction in antenna size while maintaining the original resonant frequency. This miniaturization methodology achieves compact antenna design without compromising electromagnetic performance. Compared to the non-miniaturized counterpart, the optimized configuration significantly enhances strain sensing accuracy due to its reduced footprint and refined current distribution.

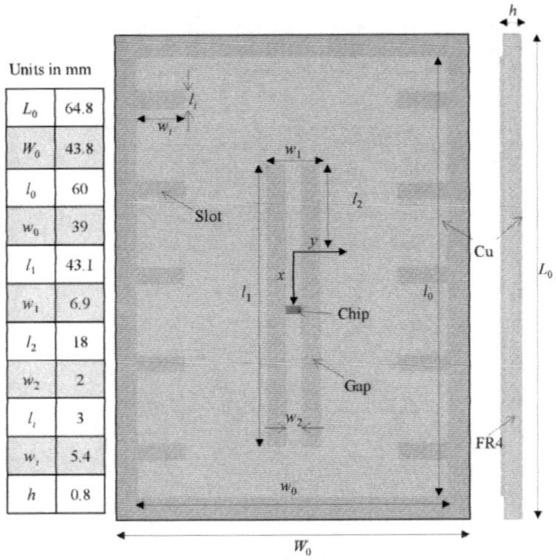

Figure 6. Optimized antenna and dimensions

3.2 Antenna mechatronic electromagnetic simulation

To simulate the resonant frequency shift of the tag antenna under external mechanical stress, this study employs COMSOL Multiphysics to implement a mechanical-electromagnetic coupled simulation. The analysis involves two sequential phases: mechanical deformation followed by electromagnetic frequency-domain computation. First, incremental displacements of 0, 500 microstrain (με), 1000 με, 1500 με, and 2000 με are applied along the antenna's longitudinal axis, with a moving mesh field dynamically updating the geometry of both the antenna and its far-field region. Subsequently, the deformed configurations are imported into the electromagnetic solver to perform a frequency-domain sweep, enabling precise extraction of the resonant frequency under each strain condition. This integrated approach rigorously quantifies strain-induced frequency shifts while maintaining multiphysics consistency, thereby establishing a robust framework for evaluating antenna performance under mechanical loading.

The simulation data acquired from COMSOL Multiphysics were post-processed to extract the return loss curves of the antenna under varying applied strain conditions, as illustrated in the figure below. This analysis quantifies the strain-dependent variations in electromagnetic performance, revealing the correlation between mechanical deformation and resonant frequency shifts through systematic characterization of the antenna's reflection coefficient (S11) across predefined microstrain levels.

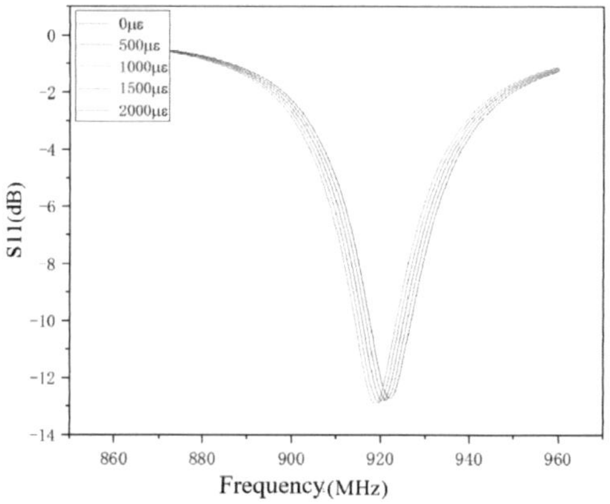

Figure 7. Reflection loss curve under different strain effects

As illustrated in the preceding figure, the resonant frequency of the tag antenna decreases monotonically with increasing axial tensile strain, which aligns with the theoretical framework established earlier. To quantify this relationship, a curve fitting analysis was performed to correlate the observed frequency shift with the applied strain, yielding the frequency shift versus strain curve shown below.

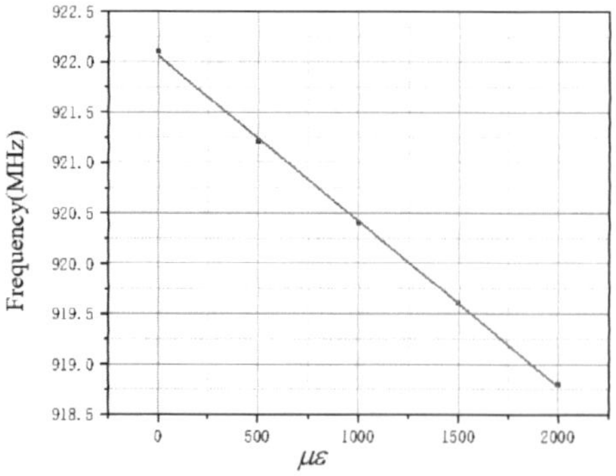

Figure 8. Fitted curve of the relationship between resonant frequency and strain

As evidenced by the figure above, the frequency shift of the antenna's resonant frequency exhibits a linear dependence on the applied strain, with an absolute slope magnitude (i.e., strain sensitivity) of

1.64 kHz/με. This result demonstrates the feasibility of employing RFID tag antennas for structural strain monitoring. Furthermore, the absence of shorting vias or similar complex structures in the proposed antenna design significantly reduces manufacturing complexity and cost, positioning this methodology as a practical solution for large-scale sensing applications.

4. Summary and prospects

This study presents a passive strain sensor utilizing an RFID patch antenna, demonstrating through theoretical modeling and electromagnetic simulations a linear correlation between resonant frequency shifts and mechanical strain ($\Delta f = -K\varepsilon$). By integrating serrated slots for geometric optimization, the sensor achieves 48% miniaturization while maintaining 1.64 kHz/με sensitivity at 925 MHz, and eliminates reliance on shorting vias to enhance manufacturing feasibility. These advancements position RFID technology as a scalable, low-cost solution for wireless structural health monitoring, extending its functionality beyond conventional identification tasks.

While simulations validate the sensor's feasibility, experimental verification is crucial to address practical limitations. Substrate creep (FR4), fabrication-induced geometric deviations, environmental impacts on dielectric permittivity, and strain transfer efficiency on metallic surfaces require empirical characterization. Experimental studies will focus on strain calibration through mechanical loading with vector network analysis, environmental cross-sensitivity assessment using controlled temperature-humidity chambers, and field validation against resistive strain gauges. These studies will refine theoretical models, optimize encapsulation protocols, and validate deployment strategies for real-world applications.

Acknowledgments

The presented work is supported by the Aviation Science Foundation of China (No. ASFC-2024Z009109001), and National Natural Science Foundation of China (No. 12002283).

Reference

[1] Wang Y, Xu Y, Zhang P, et al. Structural surface temperature measurement method based on fiber grating-Farber cavity sensor[J]. Chinese Journal of Scientific Instrument, 2022, 43(11): 40-52.
[2] Zhang M, Liu Z, Shen C, et al. A review of radio frequency identification sensing systems for structural health monitoring[J]. Materials, 2022, 15(21): 7851.
[3] Yan X, Tan Q, Simulation and design of acoustic surface wave strain sensor with high sensitivity[J]. Instrument Technique and Sensor, 2022, 59(2): 21-24
[4] Tata U, Huang H, Carter R L, et al. Exploiting a patch antenna for strain measurements[J]. Measurement Science and Technology, 2008, 20(1): 015201.
[5] Occhiuzzi C, Paggi C, Marrocco G. Passive RFID strain-sensor based on meander-line antennas[J]. IEEE Transactions on Antennas and Propagation, 2011, 59(12): 4836-4840.
[6] Occhiuzzi C, Marrocco G. Constrained-design of passive UHF RFID sensor antennas[J]. IEEE Transactions on Antennas and Propagation, 2013, 61(6): 2972-2980.
[7] Cook B S, Vyas R, Kim S, et al. RFID-based sensors for zero-power autonomous wireless sensor networks[J]. IEEE Sensors Journal, 2014, 14(8): 2419-2431.
[8] Sanders J W, Yao J, Huang H. Microstrip patch antenna temperature sensor[J]. IEEE sensors journal, 2015, 15(9): 5312-5319.
[9] Zhang J, Li C, Gao Y, et al. Flexible multimode antenna sensor with strain and humidity sensing capability for structural health monitoring[J]. Sensors and Actuators A: Physical, 2022, 347: 113960.
[10] Xie X, Dong L, Wang H, et al. Passive wireless RFID strain sensor for directional-independent structural deformation monitoring[J]. Sensors and Actuators A: Physical, 2024, 380: 116025.
[11] Cheng X. Research on design and detection method of metal component surface strain RFID sensor tag[D]. Sichuan: University of Electronic Science and Technology of China, 2022.

AUTHOR INDEX